Haskell Design Patterns

Take your Haskell and functional programming
skills to the next level by exploring new idioms
and design patterns

Ryan Lemmer

[PACKT] open source *
PUBLISHING   community experience distilled

BIRMINGHAM - MUMBAI

Haskell Design Patterns

First published: October 2015

Production reference: 1261015

Published by Packt Publishing Ltd.
Livery Place
35 Livery Street
Birmingham B3 2PB, UK.

ISBN 978-1-78398-872-3

www.packtpub.com

Credits

Author
Ryan Lemmer

Reviewer
Samuli Thomasson

Commissioning Editor
Pramila Balan

Acquisition Editor
Sonali Vernekar

Content Development Editor
Anish Dhurat

Technical Editor
Vivek Pala

Copy Editor
Pranjali Chury

Project Coordinator
Bijal Patel

Proofreader
Safis Editing

Indexer
Monica Ajmera Mehta

Graphics
Disha Haria

Production Coordinator
Arvindkumar Gupta

Cover Work
Arvindkumar Gupta

About the Author

Ryan Lemmer is software maker, coach, and strategic advisor based in Cape Town.

With a background in mathematics and computer science and 20 years of developing software in the trenches, Ryan remains inspired and humbled by the process of creating and evolving software.

Ryan is a polyglot programmer, who prefers to think in Haskell.

He loves to learn and facilitate learning for others.

This book is the story of the great work of many people. I would like to thank and acknowledge all of them here. You will meet them in this book, through their work.

Also, thanks to my editors at Packt Publishing for their support: Sonali Vernekar, Ajinkya Paranjape, and Anish Dhurat.

Thanks Lynn and Dain for all the support, and finally, thank you Christine, for everything.

About the Reviewer

Samuli Thomasson is a keen Haskell programmer who is pursuing an MSc in computer science at the University of Helsinki, Finland. He is interested in novel functional programming paradigms, distributed systems, and advanced type-safe programming methods. He likes reading books on philosophy.

He is very excited about the book *Haskell Design Patterns* by Ryan Lemmer. He recommends it to all who have a basic knowledge of pure functional programming in place and would like to get familiar with some of the more advanced and compelling features Haskell has to offer.

He enjoys sauna evenings and playing board games with his friends. You can check out his public repositories on Github at `https://github.com/SimSaladin`.

www.PacktPub.com

Support files, eBooks, discount offers, and more

For support files and downloads related to your book, please visit www.PacktPub.com.

Did you know that Packt offers eBook versions of every book published, with PDF and ePub files available? You can upgrade to the eBook version at www.PacktPub.com and as a print book customer, you are entitled to a discount on the eBook copy. Get in touch with us at service@packtpub.com for more details.

At www.PacktPub.com, you can also read a collection of free technical articles, sign up for a range of free newsletters and receive exclusive discounts and offers on Packt books and eBooks.

https://www2.packtpub.com/books/subscription/packtlib

Do you need instant solutions to your IT questions? PacktLib is Packt's online digital book library. Here, you can search, access, and read Packt's entire library of books.

Why subscribe?

- Fully searchable across every book published by Packt
- Copy and paste, print, and bookmark content
- On demand and accessible via a web browser

Free access for Packt account holders

If you have an account with Packt at www.PacktPub.com, you can use this to access PacktLib today and view 9 entirely free books. Simply use your login credentials for immediate access.

Table of Contents

Preface

This book is not a blow-by-blow translation of the Gang of Four design patterns (distilled out of the object-oriented programming paradigm). Having said that, wherever there is an intersection with Gang of Four patterns, we explore it more deeply.

This book is also not intended as a definitive taxonomy of patterns in functional programming or Haskell. Instead, this book is the story of modern Haskell, one pattern at a time, one line of code at a time. By following the historical arc of development, we can place the elements of modern Haskell in a conceptual framework more easily.

What this book covers

Chapter 1, *Functional Patterns – the Building Blocks*, explores the three pillars of Haskell, that is, first-class functions, lazy evaluation, and the Haskell type system, through the lens of patterns and idioms. We will cover some Gang of Four OOP design patterns along the way.

Chapter 2, *Patterns for I/O*, explores three ways of streaming I/O, that is, imperative, lazy, and iteratee based. While you're at it, learn about the problem with lazy I/O, together with a solution.

Chapter 3, *Patterns for Composition*, traces the hierarchy formed by functor, applicative, arrow, and monad, with a focus on how these types compose. Synthesize functor, applicative, arrow, and monad in a single conceptual framework.

Chapter 4, *Patterns of Folding and Traversing*, demonstrates how fold and map are generalized by Foldable and Traversable, which in turn are unified in a broader context by functional lenses.

Chapter 5, *Patterns of Type Abstraction*, retraces the evolution of the Haskell type system, one key language extension at a time. We'll explore RankNtypes, existensial types, phantom types, GADTs, the type-case pattern, dynamic types, heterogeneous lists, multiparameter typeclasses, and functional dependencies.

Chapter 6, *Patterns of Generic Programming*, delves into patterns of generic programming, with a focus on datatype generic programming. We will taste three flavors of generic programming: sum of products generic programming, origami programming, and scrap your boilerplate.

Chapter 7, *Patterns of Kind Abstraction*, delves into the Haskell kind system and related language extensions: associated types, type families, kind polymorphism, and type promotion. We'll get a sense of type-level programming and then conclude by going to the edge of Haskell: diving into dependently-typed programming.

What you need for this book

This book is written against the GHC compiler.

However, the first four chapters of this book are less dependent on any particular compiler. Since remaining chapters rely more heavily on the GHC language extensions, the GHC compiler matters more for those chapters.

Who this book is for

If you're a Haskell programmer with a firm grasp of the basics and ready to move more deeply into modern idiomatic Haskell programming, this book is for you.

Conventions

In this book, you will find a number of text styles that distinguish between different kinds of information. Here are some examples of these styles and an explanation of their meaning.

Code words in text are shown as follows: "We refactor `evalTry` function by using the bind operator (`>>=`)"

A block of code is set as follows:

```
data Choice a b = L a | R b
  deriving (Show)

data Combo a b = Combo a b
  deriving (Show)
```

When we wish to draw your attention to a particular part of a code block, the relevant lines or items are set in bold:

```
data Choice a b = L a | R b
    deriving (Show)

data Combo a b = Combo a b
    deriving (Show)
```

New terms and **important words** are shown in bold.

 Warnings or important notes appear in a box like this.

Reader feedback

Feedback from our readers is always welcome. Let us know what you think about this book—what you liked or disliked. Reader feedback is important for us as it helps us develop titles that you will really get the most out of.

To send us general feedback, simply e-mail feedback@packtpub.com, and mention the book's title in the subject of your message.

If there is a topic that you have expertise in and you are interested in either writing or contributing to a book, see our author guide at www.packtpub.com/authors.

Customer support

Now that you are the proud owner of a Packt book, we have a number of things to help you to get the most from your purchase.

Downloading the example code

You can download the example code files from your account at http://www.packtpub.com for all the Packt Publishing books you have purchased. If you purchased this book elsewhere, you can visit http://www.packtpub.com/support and register to have the files e-mailed directly to you.

Errata

Although we have taken every care to ensure the accuracy of our content, mistakes do happen. If you find a mistake in one of our books—maybe a mistake in the text or the code—we would be grateful if you could report this to us. By doing so, you can save other readers from frustration and help us improve subsequent versions of this book. If you find any errata, please report them by visiting http://www.packtpub. com/submit-errata, selecting your book, clicking on the **Errata Submission Form** link, and entering the details of your errata. Once your errata are verified, your submission will be accepted and the errata will be uploaded to our website or added to any list of existing errata under the Errata section of that title.

To view the previously submitted errata, go to https://www.packtpub.com/books/ content/support and enter the name of the book in the search field. The required information will appear under the **Errata** section.

Piracy

Piracy of copyrighted material on the Internet is an ongoing problem across all media. At Packt, we take the protection of our copyright and licenses very seriously. If you come across any illegal copies of our works in any form on the Internet, please provide us with the location address or website name immediately so that we can pursue a remedy.

Please contact us at copyright@packtpub.com with a link to the suspected pirated material.

We appreciate your help in protecting our authors and our ability to bring you valuable content.

Questions

If you have a problem with any aspect of this book, you can contact us at questions@packtpub.com, and we will do our best to address the problem.

1
Functional Patterns – the Building Blocks

Software design patterns were forged at a time when **object oriented programming (OOP)** reigned. This led to "design patterns" becoming somewhat synonymous with "OOP design patterns". But design patterns are solutions to problems, and "problems" are relative to the strengths and weaknesses of the context in which they occur. A design problem in OOP is not necessarily one in **functional programming (FP)**, and vice versa.

From a Haskell perspective, many (but not all) of the well known "Gang of Four" patterns [*Design patterns, Gamma et al.*] become so easy to solve that it is not worth going to the trouble of treating them as patterns. However, design patterns remain relevant for Haskell.

> "After al, as Erich Gamma said, "deja vu is language neutral"

> *Modularity means more than modules. Our ability to de-compose a problem into parts depends directly on our ability to glue solutions together. To support modular programming, a language must provide good glue."*

> *- Why Functional Programming Matters - John Hughes*

In order to have a meaningful conversation about Haskell design patterns, we'll begin our exploration by looking at the three primary kinds of "glue" in Haskell: first-class functions, the Haskell type system, and lazy evaluation. This chapter revisits the Haskell you already know through the lens of design patterns, and looks at:

- Higher-order functions
- Currying
- Recursion
- Types, pattern matching, polymorphism
- Lazy Evaluation
- Monads

Higher-order functions

Functions are our first kind of "glue" in Haskell.

Functions as first-class citizens

Haskell functions are first-class citizens of the language. This means that:

- We can name a function just as we can name any primitive value:

  ```
  square = \x -> x * x
  ```

- We can pass functions to other functions:

  ```
  map square [1, 3, 5, 7]
  ```

  (Here, `map` is a higher-order function.)

- Functions can produce other functions (here, by currying the `foldr` function):

  ```
  sum = foldr (+) 0
  ```

- Functions can form part of other data structures:

  ```
  let fs = [(* 2), (* 3), (* 5)]
  zipWith (\f v -> f v) fs [1, 3, 5]
  ```

This places Haskell functions on an equal footing with primitive types.

Composing functions

Let's compose these three functions, f, g, and h, in a few different ways:

```
f, g, h :: String -> String
```

The most rudimentary way of combining them is through nesting:

```
z x = f (g (h x))
```

Function composition gives us a more idiomatic way of combining functions:

```
z' x = (f . g . h) x
```

Finally, we can abandon any reference to arguments:

```
z'' = f . g . h
```

This leaves us with an expression consisting of only functions. This is the "point-free" form.

Programming with functions in this style, free of arguments, is called **tacit programming**.

It is hard to argue against the elegance of this style, but in practice, point-free style can be more fun to write than to read: it can become difficult to infer types (and, therefore, meaning). Use this style when ease of reading is not overly compromised.

Currying functions

Haskell allows for both curried and uncurried functions:

```
greetCurried :: String -> String -> String
greetCurried title name
  = "Greetings " ++ title ++ " " ++ name

greetUncurried :: (String, String) -> String
greetUncurried (title, name)
  = "Greetings " ++ title ++ " " ++ name
```

Let's suppose that we need a function with the first argument fixed:

```
greetCurried' :: String -> String
greetCurried' = greetCurried "Ms"

greetUncurried' :: String -> String
greetUncurried' name = greetUncurried ("Ms", name)
```

In both cases, we have applied one of the arguments and thereby specialized our original function. For the uncurried function we needed to mention all parameters in the reshaped function, while for the curried one we could just ignore subsequent arguments.

Since it is fairly easy to translate a curried function to an uncurried function (and vice versa) a question arises: why and when would one want to use uncurried functions?

Currying and composability

Consider a function that returns a few pieces of data, which you choose to express as a tuple:

```
g n = (n^2, n^3)
```

Then suppose we want to find the maximum value in that tuple:

```
max (g 11)
```

This would not work because max value is curried, but we can easily align the types by uncurrying:

```
uncurry max (g 11)
```

Whenever we have a function returning a tuple and we want to consume that tuple from a curried function, we need to uncurry that function. Alternatively, if we are writing a function to consume an output tuple from another function, we might choose to write our function in uncurried (tuple arguments) form so that we don't have to later uncurry our function or unpack the tuple.

It is idiomatic in Haskell to curry by default. There is a very important reason for this, as you will see with this example. Thanks to currying, we can do this:

```
map (map square) [[1], [2,2], [3,3,3]]
```

We cannot, however, do this:

```
let map' = uncurry map
map' (map' square) [[1], [2,2], [3,3,3]]
```

We need to explicitly curry map' in order to compose it with other functions:

```
(curry map') (curry map' square) [[1], [2,2], [3,3,3]]
```

Curried functions are composable, whereas uncurried functions are not.

Decoupling with currying

If we can apply one function argument at a time, nothing stops us from doing so at entirely different places in our codebase. For instance, we might "wire in" some authentication-related information into our function at one end of the codebase and use the specialized function in another part of the codebase that has no cognizance of the authentication argument and its related types.

This can be a powerful tool for decoupling, the site of decoupling being the function argument list!

Recursion

Recursion is even more fundamental than functions and types, in the sense that we can have recursive functions and types. Moreover, "recursion" can refer to syntax (a function or type referring to itself) or to the execution process.

Non-tail recursion

Recursion can be viewed as a pattern for avoiding a mutable state:

```
sumNonTail [] = 0
sumNonTail (x:xs) = x + (sumNonTail xs)
```

Without recursion, we would need to iterate through the list and keep adding to an intermediary sum until the list is exhausted, as shown in the following code:

```
sumNonTail [2, 3, 5, 7]
```

This first expands into a nested chain of deferred operations, and when there are only primitives left in the expression, the computation starts folding back in on itself:

```
-- 2 + sumNonTail [3, 5, 7]
-- 2 + (3 + sumNonTail [5, 7])
-- 2 + (3 + (5 + sumNonTail [7]))
-- 2 + (3 + (5 + (7 + sumNonTail [])))
-- 2 + (3 + (5 + (7 + 0)))
-- 2 + (3 + (5 + 7))
-- 2 + (3 + 12)
-- 2 + 15
-- 17
```

The sumNonTail function is non-tail-recursive. Because the recursion is "trapped" by the + operator, we need to hold the entire list in memory to perform the sum.

Tail recursion

Tail recursion addresses the exorbitant use of space we have with non-tail-recursive processes:

```
sumTail' acc [] = acc
sumTail' acc (x:xs) = sumTail' (acc + x) xs
sumTail xs = sumTail' 0 xs
```

This form of recursion looks less like mathematical induction than the sumNonTail function did, and it also requires a helper function sumTail' to get the same ease of use that we had with sumNonTail. The advantage is clear when we look at the use of "constant space" in this process:

```
-- sumTail [2, 3, 5, 7]
-- sumTail' 0 [2, 3, 5, 7]
-- sumTail' 2 [3, 5, 7]
-- sumTail' 5 [5, 7]
-- sumTail' 10 [7]
-- sumTail' 17 []
-- 17
```

Even though sumTail is a recursive function, it expresses an iterative process. sumNonTail is a recursive function that expresses a recursive process.

Folding abstracts recursion

Tail recursion is captured by the foldl function, as shown in the following code:

```
foldlSum = foldl (+) 0
```

The foldl function expands in exactly the same way as sumTail'. In contrast, foldrSum expands in the same way as sumNonTail:

```
foldrSum = foldr (+) 0
```

One can clearly see the tail recursion in the definition of foldl, whereas in the definition of foldr, recursion is "trapped" by f:

```
foldr _ v [] = v
foldr f v (x:xs) = f x (foldr f v xs)

foldl _ v [] = v
foldl f v (x:xs) = foldl f (f v x) xs
```

Types, pattern matching, and polymorphism

Algebraic types give us a very concise way to model composite types, even recursive ones. Pattern matching makes it easy to work with algebraic types. Type classes enable both the fundamental types of polymorphism: parametric and ad-hoc.

Together, these capabilities allow us to easily express many of the Gang of Four patterns.

Algebraic types and pattern matching

Algebraic data types can express a combination of types, for example:

```
type Name = String
type Age = Int
data Person = P String Int -- combination
```

They can also express a composite of alternatives:

```
data MaybeInt = NoInt | JustInt Int
```

Here, each alternative represents a valid constructor of the algebraic type:

```
maybeInts = [JustInt 2, JustInt 3, JustInt 5, NoInt]
```

Type combination is also known as "product of types" and type alternation as "sum of types". In this way, we can create an "algebra of types", with sum and product as operators, hence the name **Algebraic data types**.

By parameterizing algebraic types, we create generic types:

```
data Maybe' a = Nothing' | Just' a
```

Algebraic data type constructors also serve as "deconstructors" in pattern matching:

```
fMaybe f (Just' x) = Just' (f x)
fMaybe f Nothing' = Nothing'

fMaybes = map (fMaybe (* 2)) [Just' 2, Just' 3, Nothing']
```

On the left of the = sign we deconstruct; on the right, we construct. In this sense, pattern matching is the complement of algebraic data types: they are two sides of the same coin.

Recursive types

We capture the "composite pattern" very succinctly by creating recursive algebraic types, for example:

```
data Tree a = Leaf a | Branch (Tree a) (Tree a)
```

This pattern describes the need to sometimes unify a composite structure with individual members of that structure. In this case, we're unifying `Leaf` (a leaf being a part of a tree) and `Tree` (the composite structure). Now we can write functions that act on trees and leaves:

```
size :: Tree a -> Int
size (Leaf x) = 1
size (Branch t u) = size t + size u + 1
```

Functions over recursive types are typically recursive themselves.

Polymorphism

Polymorphism points at the phenomenon of something taking many forms.
In Haskell, there are two kinds of polymorphism: parametric and ad-hoc (first described by Strachey in *Fundamental Concepts in Programming Languages*, 1967).

Parametric polymorphism

In the following code, we have a function defined on a list of any type. The function is defined at such a high level of abstraction that the precise input type simply never comes into play, yet the result is of a particular type:

```
length' :: [a] -> Int
length' [] = 0
length' (x:xs) = 1 + length xs
```

The `length` object exhibits parametric polymorphism because it acts uniformly on a range of types that share a common structure, in this case, lists:

```
length' [1,2,3,5,7]
length' ['1','2','3','5','7']
```

In this sense, `length` is a generic function. Functions defined on parametric data types tend to be generic.

Ad-hoc polymorphism

"Wadler conceived of type classes in a conversation with Joe Fasel. Fasel had in mind a different idea, but it was he who had the key insight that overloading should be reflected in the type of the function. Wadler misunderstood what Fasel had in mind, and type classes were born!"

- History of Haskell, Hudak et al.

The canonical example of ad-hoc polymorphism (also known as "overloading") is that of the polymorphic + operator, defined for all Num variables:

```
class Num a where
    (+) :: a -> a -> a

instance Int Num where
    (+) :: Int -> Int -> Int
    x + y = intPlus x y

instance Float Num where
    (+) :: Float -> Float -> Float
    x + y = floatPlus x y
```

In fact, the introduction of type classes into Haskell was driven by the need to solve the problem of overloading numerical operators and equality.

When we call (+) on two numbers, the compiler will dispatch evaluation to the concrete implementation, based on the types of numbers being added:

```
let x_int = 1 + 1        -- dispatch to 'intPlus'
let x_float = 1.0 + 2.5  -- dispatch to 'floatPlus'
let x   = 1 + 3.14       -- dispatch to 'floatPlus'
```

In the last line, we are adding what looks like an int to a float variable. In many languages, we'd have to resort to explicit **coercion** (of int to float, say) to resolve this type of "mismatch". In Haskell, this is resolved by treating the value of 1 as a type-class **polymorphic value**:

```
ghci> :t 1 -- Num a => a
```

1 is a generic value (a Num variable); whether 1 is an int variable or a float variable (or a fractional, say) depends on the context in which it will appear.

Alternation-based ad-hoc polymorphism

There are two kinds of ad-hoc polymorphism. We've seen the first type already in this chapter:

```
data Maybe' a = Nothing' | Just' a
fMaybe f (Just' x) = Just' (f x)
fMaybe f Nothing' = Nothing'
```

The `fMaybe` function is polymorphically defined over the alternations of `Maybe`. In order to directly contrast the two kinds of polymorphism, let's carry this idea over into another example:

```
data Shape = Circle Float | Rect Float Float

area :: Shape -> Float
area (Circle r) = pi * r^2
area (Rect length width) = length * width
```

The `area` function is dispatched over the alternations of the `Shape` type.

Class-based ad-hoc polymorphism

We could also have achieved a polymorphic `area` function over shapes in this way:

```
data Circle = Circle Float
data Rect = Rect Float Float

class Shape a where
  area :: a -> Float

instance Shape Circle where
  area (Circle r) = pi * r^2
instance Shape Rect where
  area (Rect length' width') = length' * width'
```

> **Downloading the example code**
>
> You can download the example code files from your account at http://www.packtpub.com for all the Packt Publishing books you have purchased. If you purchased this book elsewhere, you can visit http://www.packtpub.com/support and register to have the files e-mailed directly to you.

Instead of unifying shapes with an algebraic "sum of types", we created two distinct shape types and unified them through a `Shape` class. This time the `area` function exhibits class-based polymorphism.

Alternation-based versus class-based

It is tempting to ask "which approach is best?" Instead, let's explore the important ways in which they differ:

	Alternation-based	Class-based
Different coupling between function and type	The function type refers to the algebraic type Shape and then defines special cases for each alternative.	The function type is only aware of the type it is acting on, not the Shape "super type".
Distribution of function definition	The overloaded functions are defined together in one place for all alternations.	Overloaded functions all appear in their respective class implementations. This means a function can be overloaded in very diverse parts of the codebase if need be.
Adding new types	Adding a new alternative to the algebraic type requires changing all existing functions acting directly on the algebraic "super type"	We can add a new type that implements the type class without changing any code in place (only adding). This is very important since it enables us to extend third-party code.
Adding new functions	A perimeter function acting on Shape won't be explicitly related to area in any way.	A perimeter function could be explicitly related to area by adding it to the Shape class. This is a powerful way of grouping functions together.
Type expressivity	This is useful for expressing simple type hierarchies.	We can have multiple, orthogonal hierarchies, each implementing the type class (For example, we can express multiple-inheritance type relations). This allows for modeling much richer data types.

Polymorphic dispatch and the visitor pattern

While exploring ad-hoc polymorphism, we saw how we can simulate static type dispatch ("static" meaning that the dispatch is resolved at compile time, as opposed to "dynamic dispatch", resolved only at runtime). Let's return to our area function:

```
area (Circle 10)
```

The preceding command will dispatch to the overloaded `area` function by matching:

- A sub type of the `Shape` algebraic type (subtype-based)
- The type class to which `Circle` belongs that is, `Shape` (class-based)

We've referred to this as "dispatching on type" but, strictly speaking, type dispatch would have to resemble the following invalid Haskell:

```
f v = case (type v) of
   Int -> "Int: " ++ (show v)
   Bool -> "Bool" ++ (show v)
```

Having said that, pattern-based and type-based dispatching are not that far apart:

```
data TypeIntBool = Int' Int | Bool' Bool

f :: TypeIntBool -> String
f (Int' v) = "Int: " ++ (show v)
f (Bool' v) = "Bool: " ++ (show v)
```

So far, we have only seen dispatching on one argument or "single dispatch". Let's explore what "double-dispatch" might look like:

```
data CustomerEvent = InvoicePaid Float | InvoiceNonPayment
data Customer = Individual Int | Organisation Int

payment_handler :: CustomerEvent -> Customer -> String

payment_handler (InvoicePaid amt) (Individual custId)
   = "SendReceipt for " ++ (show amt)
payment_handler (InvoicePaid amount) (Organisation custId)
   = "SendReceipt for " ++ (show amt)

payment_handler InvoiceNonPayment (Individual custId)
   = "CancelService for " ++ (show custId)
payment_handler InvoiceNonPayment (Organisation custId)
   = "SendWarning for " ++ (show custId)
```

The `payment_handler` object defines behavior for all four permutations of `CustomerEvent` and `Customer`. In an OOP language, we would have to resort to the visitor pattern to achieve multiple dispatch.

> *"Visitor lets you define a new operation without changing the classes of the elements on which it operates...*
>
> *Languages that support double or multiple dispatch lessen the need for the Visitor pattern."* - Design Patterns, Gamma et al.

Unifying parametric and ad-hoc polymorphism

On the one hand, we have parametric polymorphism, where a single generic function acts on a variety of types. This is in contrast to ad-hoc polymorphism, where we have an overloaded function that is resolved to a particular function by the compiler. Put another way, parametric polymorphism allows us to lift the level of abstraction, whereas ad-hoc polymorphism gives us a powerful tool for decoupling.

In this sense, parametric polymorphism is considered to be "true polymorphism", while ad hoc is only "apparent" (or "synthetic").

Haskell blurs the distinction between ad hoc (specialized) and parametric (generic) polymorphism. We can see this clearly in the definition of the type class for equality:

```
class Eq a where
  (==), (/=) :: a -> a -> Bool
  x == y = not (x /= y)
  x /= y = not (x == y)
```

(==) and (/=) are both given mutually recursive default implementations. An implementer of the `Eq` class would have to implement at least one of these functions; in other words, one function would be specialized (ad-hoc polymorphism), leaving the other defined at a generic level (parametric polymorphism). This is a remarkable unification of two very different concepts.

Functions, types, and patterns

Functions and types intersect in several ways. Functions have a type, they can act on algebraic types, they can belong to type classes, and they can be specific or generic in type. With these capabilities, we can express several more Gang of Four patterns.

The strategy pattern

Thanks to **higher-order functions (HOF)**, we can easily inject behavior:

```
strategy fSetup fTeardown
  = do
     setup
     -- fullfil this function's purpose
     teardown
```

Here, we are defining an abstract algorithm by letting the caller pass in functions as arguments, functions that complete the detail of our algorithm. This corresponds to the strategy pattern, also concerned with decoupling an algorithm from the parts that may change.

> *"Strategy lets the algorithm vary independently from clients that use it."*

> *- Design Patterns, Gamma et al.*

The template pattern

In "OOP speak", the strategy pattern uses delegation to vary an algorithm, while the template pattern uses inheritance to vary parts of an algorithm. In Haskell, we don't have OOP inheritance, but we have something far more powerful: type classes. We might easily abstract an algorithm with this type class that acts as an abstract class:

```
class TemplateAlgorithm where
  setup :: IO a → a
  teardown :: IO a → a
  doWork :: a → a
  fulfillPurpose
    = do
       setup
       doWork
       teardown
```

> *"Define the skeleton of an algorithm in an operation, deferring some steps to subclasses. Template Method lets subclasses redefine certain steps of an algorithm without changing the algorithm's structure."*

> *- Design Patterns, Gamma et al.*

The iterator pattern

> *"Provide a way to access the elements of an aggregate object sequentially without exposing its underlying representation"*

> *- Design Patterns, Gamma et al.*

The `map` function takes care of navigating the structure of the list, while the `square` function only deals with each element of the list:

```
map square [2, 3, 5, 7]
```

We have decoupled flow control from function application, which is akin to the iterator pattern.

Decoupling behavior and modularizing code

Whenever we pass one function into another, we are decoupling two parts of code. Besides allowing us to vary the different parts at different rates, we can also put the different parts in different modules, libraries, or whatever we like.

Lazy evaluation

The history of Haskell is deeply entwined with the history of lazy evaluation.

> *"Laziness was undoubtedly the single theme that united the various groups that contributed to Haskell's design...*

> *Once we were committed to a lazy language, a pure one was inescapable."*

> *- History of Haskell, Hudak et al*

Thanks to lazy evaluation, we can still consume the undoomed part of this list:

```
doomedList = [2, 3, 5, 7, undefined]
take 0 xs = []
take n (x:xs) = x : (take (n-1) xs)

main = do print (take 4 doomedList)
```

The `take` object is lazy because the cons operator (`:`) is lazy, which is because all functions in Haskell are lazy by default.

A lazy cons evaluates only its first argument, while the second argument, the tail, is only evaluated when it is selected. (For strict lists, both head and tail are evaluated at the point of construction of the list.)

The proxy pattern has several motivations, one of which is to defer evaluation; this aspect of the proxy pattern is subsumed by lazy evaluation.

Streams

The simple idea of laziness enables has the profound effect of enabling self-reference:

```
infinite42s = 42 : infinite42s
```

Streams (lazy lists) simulate infinity through "*the promise of potential infinity*" [*Why Functional Programming Matters, Hughes*]:

```
potentialBoom = (take 5 infinite42s)
```

A stream is always just one element cons'ed to a tail of whatever size. A function such as `take` consumes its input stream but is decoupled from the producer of the stream to such an extent that it doesn't matter whether the stream is finite or infinite (unbounded). Let's see this in action with a somewhat richer example:

```
generate :: StdGen -> (Int, StdGen)
generate g = random g :: (Int, StdGen)

-- import System.Random
main = do
  gen0 <- getStdGen
  let (int1, gen1) = (generate g)
  let (int2, gen2) = (generate gen1)
```

Here we are generating a random `int` value and returning a new generator, which we could use to generate a subsequent random `int` value (passing in the same generator would yield the same random number).

Carrying along the generator from one call to the next pollutes our code and makes our intent less clear. Let's instead create a producer of random integers as a stream:

```
randInts' g = (randInt, g) : (randInts' nextGen)
      where (randInt, nextGen) = (generate g)
```

Next, suppress the generator part of the stream by simply selecting the first part of the tuple:

```
randInts g = map fst (randInts' g)
main = do
  g <- getStdGen
  print (take 3 (randInts g))
```

We still pass in the initial generator to the stream, but now we can consume independently from producing the numbers. We could just as easily now derive a stream of random numbers between 0 and 100:

```
randAmounts g = map (\x -> x `mod` 100) (randInts g)
```

This is why it is said that lazy lists decouple consumers from producers. From another perspective, we have a decoupling between iteration and termination. Either way, we have decoupling, which means we have a new way to modularize and structure our code.

Modeling change with streams

Consider a banking system, where we want to record the current balance of a customer's account. In a non-pure functional language, we would typically model this with a mutable variable for the balance: for each debit and credit in the "real world", we would mutate the balance variable.

> *"Can we avoid identifying time in the computer with time in the modeled world?"*
>
> *[Structure and Interpretation of Computer Programs, p. 316],*
> *Abelson and Sussman*

Yes, we can describe the evolution of a variable as a function of time. Instead of a mutable variable for bank balance, we have a sequence of balance values. In other words, we replace a mutable variable with the entire history of states:

```
bankAccount openingB (amt:amts)
  = openingB : bankAccount (openingB + amt) amts

(take 4 (bankAccount 0 [-100, 50, 50, 1]))
```

Here we have modeled the bank account as a process, which takes in a stream of transaction amounts. In practice, amounts are more likely to be an unbounded stream, which we can easily simulate with our `randAmounts` stream from earlier:

```
(take 4 (bankAccount 0 (randAmounts g)))
```

> *"if we concentrate on the entire time history, we do not emphasize change"*
>
> *- [Structure and Interpretation of Computer Programs, p. 317],*
> *Abelson and Sussman*

Lazy evil

Streams provide an antidote to mutation, but as with all powerful medicine, streams create new problems. Because streams pretend to express a complete list while only incrementally materializing the list, we cannot know exactly when evaluation of list elements happens. In the presence of side effects, this ignorance of the order of events becomes a serious problem. We will devote the next chapter to dealing with mutation in the presence of laziness.

Monads

The monad typeclass is best understood by looking at from many perspectives. That is why this book has no definitive section or chapter on monad. Instead, we will successively peel of the layers of this abstraction and make good use of it.

Let's begin by looking at a simple example of interpreting expressions:

```
data Expr = Lit Int | Div Expr Expr

eval :: Expr -> Int
eval (Lit a) = a
eval (Div a b) = eval a `div` eval b
```

The `eval` function interprets expressions written in our `Expr` data type:

```
(eval (Lit 42))           -- 42
(eval (Div (Lit 44) (Lit 11)))   -- 4
```

Stripped of real-world concerns, this is very elegant. Now let's add (naive) capability to deal with errors in our interpreter. Instead of the `eval` function returning integers, we'll return a `Try` data type, which caters for success (`Return`) and failure (`Error`):

```
data Try a = Err String | Return a
```

The refactored `evalTry` function is now much more syntactically noisy with case statements:

```
evalTry :: Expr -> Try Int
evalTry (Lit a) = Return a
evalTry (Div a b) = case (evalTry a) of
    Err e     -> Err e
    Return a' -> case (evalTry b) of
      Err e  -> Err e
      Return b' -> divTry a' b'

-- helper function
```

```
divTry :: Int -> Int -> Try Int
divTry a b = if b == 0
    then Err "Div by Zero"
    else Return (a `div` b)
```

The reason for the noise is that we have to explicitly propagate errors. If (evalTry a) fails, we return Err and bypass evaluation of the second argument.

We've used the Try data type to make failure more explicit, but it has come at a cost. This is precisely where monads come into play. Let's make our Try data type a monad:

```
instance Monad Try where
   return x   = Return x
   fail msg   = Err msg

   Err e     >>= _    = Err e
   Return a >>= f    = f a
```

Next, we refactor evalTry by using the bind operator (>>=):

```
evalTry' :: Expr -> Try Int
evalTry' (Lit a)    = Return a
evalTry' (Div a b) = (evalTry' a) >>= \a' ->
                          (evalTry' b) >>= \b' ->
                            divTry a' b'


-- evalTry' (Div (Lit 44) (Lit 0))
```

The bind operator enables error propagation:

```
Err e   >>= _    = Err e
```

Whenever we have an Err function, the subsequent part of the bind chain will be ignored and will thereby propagate the error. While this gets rid of our case statements, it is hardly very friendly. Let's rewrite it using the do notation:

```
evalTry'' (Lit a) = Return a
evalTry'' (Div a b)
  = do
     a' <- (evalTry' a)
     b' <- (evalTry' b)
     divTry a' b'
```

The `Try` data type helped us make failure more explicit, while making it a monad made it easier to work with. In this same way, monads can be used to make many other "effects" more explicit.

> *"Being explicit about effects is extremely useful, and this is something that we believe may ultimately be seen as one of Haskell's main impacts on mainstream programming"*

> *- History of Haskell, Hudak et al.*

The IO monad is particularly interesting and played an important role in the development of Haskell. When Haskell was first conceived, there were no monads and also no clear solution to the *"problem of side effects"*. In 1989, Eugenio Moggi used monads, from Category theory, to describe programming language features. Phillip Wadler, a then member of the Haskell Committee, recognized that it was possible to express Moggi's ideas in Haskell code:

> *"Although Wadler's development of Moggi's ideas was not directed towards the question of input/output, he and others at Glasgow soon realised that monads provided an ideal framework for I/O"*

> *-- History of Haskell, Hudak et al*

Because Haskell is purely functional, side effects call for special treatment. We will devote a whole chapter to exploring this topic in *Chapter 2, Patterns for I/O*.

Composing monads and structuring programs

As useful as monads are for capturing effects, we also need to compose them, for example, how do we use monads for failure, I/O, and logging together?

In the same way that functional composition allows us to write more focused functions that can be combined together, monad composition allows us to create more focused monads, to be recombined in different ways. In *Chapter 3: Patterns for Composition*, we will explore monad transformers and how to create "monad stacks" of transformers to achieve monad composition.

Summary

In this chapter, we explored the three primary kinds of "glues" that Haskell provides: functions, the type system, and lazy evaluation. We did so by focusing on composability of these building blocks and found that wherever we can compose, we are able to decompose, decouple, and modularize our code.

We also looked at the two main kinds of polymorphism (parametric and ad hoc) as they occur in Haskell.

This chapter set the scene for starting our study of design patterns for purely functional programming in Haskell.

The next chapter will focus on patterns for I/O. Working on I/O in the face of lazy evaluation is a minefield, and we would do well with some patterns to guide us.

2
Patterns for I/O

"I believe that the monadic approach to programming, in which actions are first class values, is itself interesting, beautiful and modular. In short, Haskell is the world's finest imperative programming language"

- Tackling the Awkward Squad, Simon Peyton Jones

It is remarkable that we can do side-effecting I/O in a pure functional language!

We start this chapter by establishing I/O as a first-class citizen of Haskell. A the bulk of this chapter is concerned with exploring three styles of I/O programming in Haskell.

We start with the most naïve style: imperative style. From there, we move on to the elegant and concise "lazy I/O", only to run into its severe limitations. The way out is the third and last style we explore: iteratee I/O.

As a binding thread, we carry a simple I/O example through all three styles. We will cover the following topics:

- I/O as a first-class citizen
- Imperative I/O
- Lazy I/O
- The problem with Lazy I/O
- Resource management with bracket
- Iteratee I/O

I/O as a first class citizen

The **IO monad** provides the context in which the side effects may occur, and it also allows us to decouple pure code from the I/O code. In this way, side effects are isolated and made explicit. Let's explore the ways in which I/O participates as a first-class citizen of the language:

```
import System.IO
import Control.Monad
import Control.Applicative

main = do
  h <- openFile "jabberwocky.txt" ReadMode
  line  <- hGetLine h
  putStrLn . show . words $ line
  hClose h
```

This code looks imperative in style: it seems as if we are assigning values to the h and line objects, reading from a file, and then leaving side effects with the putStrLn function.

The openFile and hGetLine functions are I/O actions that return a file handle and string, respectively:

```
openFile :: FilePath -> IOMode -> IO Handle
hGetLine ::             Handle -> IO String
```

The hClose and putStrLn functions are I/O actions that return nothing in particular:

```
putStrLn :: String -> IO ()
hClose   :: Handle -> IO ()
```

In the putStrLn . show function, we compose a function that returns an I/O action with a pure function:

```
(putStrLn :: String -> IO ()) .
  (show :: Show a => a -> String)

putStrLn . show :: Show a => a -> IO ()
```

From this, we can see that functions can return I/O actions; functions can take I/O actions as arguments. We can compose regular functions with functions that return I/O actions. This is why it is said that I/O is a first-class citizen of Haskell. Moreover, monad implements several type classes, each of which represents a substyle of I/O programming.

Although `do` is syntactic sugar for `bind`, in a case like this, where there is just one "bind", using the bind notation is more concise:

```
hGetLine h >>= print . words
-- vs
--   line  <- hGetLine h
--   print . words $ line
```

(where *print = putStrLn . show*)

I/O as a functor, applicative, and monad

So far, we have described I/O as a monad. In the next chapter, we will do an in-depth survey of the hierarchy formed by the functor, applicative, and monad. For our purposes here, simply note that the I/O monad is also an applicative functor, which in turn is also a functor. I/O as a functor allows us to use the `fmap` function in the result of the `hGetLineh` action:

```
line <- fmap (show . words) (hGetLine h)
putStrLn line
```

I/O as an applicative functor means that we can use this syntax instead of the `fmap` function:

```
line <- (show . words) <$> (hGetLine h)
putStrLn line
```

For the monadic version of the preceding code, we use the `liftM` function:

```
line <- liftM (show . words) (hGetLine h)
putStrLn line
```

All three styles of the `fmap` function are equivalent in this case. However, as we will see, monad is more powerful than applicative, and applicative more powerful than functor. Simply put, monad enables us to compose I/O actions together in sequenced pipelines; functor and applicative allow us to apply functions to I/O actions.

Imperative I/O

Even though the Haskell I/O code is purely functional, this does not prevent us from writing imperative style I/O code. Let's start by printing all the lines of the file:

```
import System.IO
import qualified Data.ByteString as B
import qualified Data.ByteString.Char8 as B8
```

```
import Data.Char (chr)

main = do
  h <- openFile "jabberwocky.txt" ReadMode
  loop h
  hClose h
  where
    loop h' = do
     isEof <- hIsEOF h'
     if isEof
       then putStrLn "DONE..."
       else do
         line  <- hGetLine h'
         print $ words line
         loop h'
```

Instead of the hGetLine function, let's use a Data.ByteString.hGet function to read from the file in chunks of 8 bytes:

```
chunk  <- B.hGet h' 8

print . words $ show chunk
-- vs
-- line  <- hGetLine h'
-- print $ words line
```

The splitting of a chunk into words is not meaningful anymore. We need to accumulate the chunks until we reach the end of a line and then capture the accumulated line and possible remainder:

```
data Chunk = Chunk   {chunk :: String}
           | LineEnd {chunk :: String,
                      remainder :: String}
      deriving (Show)
```

If the chunk contains a newline character, then we return the LineEnd function; otherwise, it is just another Chunk function, as shown in the following code:

```
parseChunk chunk
   = if rightS == B8.pack ""
      then Chunk   (toS leftS)
      else LineEnd (toS leftS) ((toS . B8.tail) rightS)
     where
       (leftS, rightS) = B8.break (== '\n') chunk
       toS = map (chr . fromEnum) . B.unpack
```

Let's use the parseChunk function:

```
main = do
  print $ (parseChunk (B8.pack "AAA\nBB"))
  -- LineEnd {chunk = "AAA", remainder = "BB"}
  print $ (parseChunk (B8.pack "CCC"))
  -- Chunk {chunk = "CCC"}
```

Now we can accumulate chunks into lines:

```
main = do
  fileH <- openFile "jabberwocky.txt" ReadMode
  loop "" fileH
  hClose fileH
  where
  loop acc h = do
    isEof <- hIsEOF h
    if isEof
      then do putStrLn acc; putStrLn "DONE..."
      else do
        chunk <- B.hGet h 8
        case (parseChunk chunk) of
          (Chunk chunk')
-> do
    let accLine = acc ++ chunk'
    loop accLine
          h
          (LineEnd chunk' remainder)
-> do
    let line = acc ++ chunk'
    -- process line...
    putStrLn line
    loop remainder h
```

For each loop iteration, we use the hGet function, a chunk of file. When we reach a LineEnd chunk, we continue looping, this time starting with the remainder method as the first chunk in a new accumulation of a line. For regular chunks, we just add the chunk to the accumulation and continue looping.

What we call "imperative I/O", is strictly called "handle-based I/O" here. For more information, visit http://okmij.org/ftp/Haskell/Iteratee/describe.pdf.

Handle-based I/O has some good characteristics, which are as follows:

- Processing is incremental (for example, the processing of a file).
- We have precise control over resources (for example, when files are opened or closed and when long-running processes are started).

The downsides of handle-based I/O are as follows:

- I/O is expressed at a relatively low level of abstraction.

- This style of code is not very composable; for example, in the previous code, we interleave the iteration of the file with the processing of the chunks.

- We have the "exposed traversal state: we need to pass the file handle around, and check for EOF at each iteration. We need to explicitly clean up the resource.

Lazy I/O

Of the three main glues of Haskell (HOFs, the type system, and laziness), laziness is different in that it is not a concrete thing in the language, but is instead related to the way the code will be evaluated in the runtime. Laziness is something that we have to know about rather than something we can always see in the code.

Of course, laziness does show in the language, for example, wherever we want to enforce strict evaluation, as with the `sequence` function:

```
main = do
  -- lazy IO stream
  let ios = map putStrLn ["this", "won't", "run"]

  putStrLn "until ios is 'sequenced'..."
  sequence_ ios -- perform actions
```

Where:

```
sequence_  :: [IO ()] -> IO ()
sequence_  =  foldr (>>) (return ())
```

The `sequence_` function discards the action results because the `(>>)` operator discards the results.

In contrast, the `sequence` function retains the results:

```
main = do
  h <- openFile "jabberwocky.txt" ReadMode
  line1 <- hGetLine h                  -- perform action
  let getLines = [hGetLine h, hGetLine h]
  [line2, line3] <- sequence getLines -- perform actions
  hClose h
  putStrLn line2
```

The `line1` function is read eagerly with the first `hGetLine h` function, while the `line2` and `line3` functions are only read when "sequenced".

While the `hGetLine` function returns a strict string, the `hGetContents` function returns a lazy string. Put another way, the `hGetContents` streams file contents on demand while `hGetLine` function doesn't:

```
main = do
  h <- openFile "jabberwocky.txt" ReadMode
  contents  <- hGetContents h
  putStrLn (take 10 contents) -- lazily fetch 10 chars
  hClose h
```

Using the `lines` function together with `hGetContents` function, we get a lazy stream of file lines:

```
lineStream h = hGetContents h >>= return . lines
main = do
  h <- (openFile "jabberwocky.txt" ReadMode)
  lines' <- lineStream h
  sequence_ (map putStrLn lines')
  hClose h
```

The `mapM` function captures the common pattern of mapping and sequencing:

```
main = do
  h <- (openFile "jabberwocky.txt" ReadMode)
  lines' <- lineStream h
  mapM_ putStrLn lines'
  hClose h
```

Here, `mapM_ f` is equal to `sequence_ (map f)`.

The `forM_` function is just a `mapM_` function with flipped arguments, which is useful when you want to pass a "trailing lambda":

```
main = do
  h <- (openFile "jabberwocky.txt" ReadMode)
  lines' <- lineStream h
  forM_ lines' $ \line -> do
      let reversed = reverse line
      putStrLn reversed
  hClose h
```

When performing a lazy I/O, we need to make the distinction between an I/O action and performing an I/O action. Also, we need to know the lazy/strict characteristics of the functions we are working with (for example, `hGetLine` and `hGetContents`).

Let's return to our imperative style I/O code from the previous section and rephrase it in the style of lazy I/O. We'll retain the `Chunk` data type and the `parseChunk` function from the imperative example:

```
data Chunk = Chunk    {chunk :: String}
                    | LineEnd {chunk :: String,
                                    remainder :: String}


parseChunk :: String -> Chunk
-- parseChunk (B8.pack "gimble in the wabe:\nAll")
-- gives: LineEnd "gimble in the wabe"
--                          "All"
```

To write this in a lazy I/O style, we'll start by defining a stream of file chunks:

```
-- import qualified Data.ByteString.Lazy as LB
-- import qualified Data.ByteString.Lazy.Char8 as L8
chunkStream :: Handle -> IO [L8.ByteString]
chunkStream h
  = do
    isEof <- hIsEOF h
    if isEof
      then return []
      else do
        chunk <- LB.hGet h 8
        rest  <- (chunkStream h)
        return (chunk:rest)
```

Now we can produce a stream and consume it:

```
main = do
   chunks <- chunkStream h
   print $ take 10 chunks
```

The chunk stream produces data. Next, we write a consumer:

```
processChunk :: String -> [L8.ByteString] -> IO ()
processChunk acc []
   = do putStrLn acc -- terminate recursion

processChunk' acc (chunk:chunks)
   = case (parseChunk chunk) of
     (Chunk chunk')
```

```
    -> do
      processChunk' (acc ++ chunk') chunks
  (LineEnd chunk' remainder)
    -> do
      let line = acc ++ chunk'
      putStrLn line -- do something with line
      processChunk' remainder chunks

processChunk = processChunk' ""
```

The `processChunk` method recursively consumes our stream and accumulates chunks into lines. It is tail recursive and uses the constant space, as shown in the following code:

```
main = do
  h <- openFile "jabberwocky.txt" ReadMode
  chunkStream h >>= processChunk
  hClose h
```

We decoupled a producer from the consumer and the consumer drives the materializing of the source stream.

In contrast, in the imperative example, the `loop` function drives the iteration through the chunks. Also, in the imperative case, the consumer is not explicit.

The `processChunk` method is an I/O action with side effects. As it loops through the file chunks (via recursion), it keeps accumulating chunks until it has captured a whole line.

Then it does some I/O with the line `putStrLn line` and starts accumulating chunks for the next line. Iteration and I/O processing are interleaved. We can decouple this further, by making a pure function, `lineStream`, that produces a stream of lines:

```
lineStream accChunks [] = [accChunks]
lineStream accChunks (chunk:chunks)
  = case (parseChunk chunk) of
    (Chunk chunk')
      -> lineStream (accChunks ++ chunk') chunks
    (LineEnd chunk' remainder)
  -> (accChunks ++ chunk') :
        (lineStream remainder chunks)
toLines = lineStream ""
```

Now we can feed the `chunkStream` function to the `toLines` function:

```
main = do
  h <- openFile "jabberwocky.txt" ReadMode
  lines' <- liftM toLines (chunkStream h)
  mapM_ putStrLn lines'
  hClose h
```

The `toLines` is a pure function of its input stream while `chunkStream` is a stream wrapped in an I/O monad. This is why we need `liftM` to lift the toLines function into the Monad. We could have said this instead:

```
  chunks <- (chunkStream h)
  let lines' = toLines chunks
```

It is only when we do the `mapM putStrLn lines'` method that the stream `toLines` starts to materialize, which in turn drives the evaluation of `chunkStream`. When we attempt to print one line, the line is materialized, just in time, lazily. Underneath the `mapM_` function, it is the `sequence_` function that drives the evaluation of the stream.

This brings us back full circle to the `lines` library function, which lazily returns lines from file chunks:

```
main = do
  h <- openFile "jabberwocky.txt" ReadMode
  lines' <- hGetContents h >>= return . Lines
  -- vs
  -- lines' <- liftM toLines (chunkStream h)
  mapM putStrLn lines'
  hClose h
```

In the preceding code, we composed pure functional streams with I/O streams using monadic operators and functions. This is the lazy I/O, the pure functional way for composable I/O. In the same way that we can often express functions as pipelines of simpler functions, the same is true for I/O. Many practical I/Os can be modeled as processing pipelines of streams.

Let's try to describe the essence of Lazy I/O, starting with the advantages:

- I/O is expressed at a relatively high level of abstraction
- It is very composable, enabling the decoupling of producers from consumers

The disadvantages of Lazy I/O are as follows:

- It has poor control over when something is evaluated
- It has a poor control of resources

Lazy evaluation is elegant and has far reaching implications. It is an integral part of pure functional code that unfortunately does not translate to the I/O code.

The problems with lazy I/O

Let's use the `hGetLine` function alongside the `hGetContents` function:

```
main = do
  h <- openFile "jabberwocky.txt" ReadMode
  firstLine <- hGetLine h      -- returns a string
  contents  <- hGetContents h -- returns a "promise"

  hClose h              -- close file
  print $ words firstLine
  print $ words contents
```

We close the file before consuming the `firstLine` string and the `contents` stream:

```
print $ words firstLine
  ["'Twas","brillig,","and","the","slithy","toves"]
print $ words contents
  []
```

The `contents` is a live stream that gets turned off when the file is closed. The `firstLine` is an eager string and survives the closing of the file.

The preceding example points to some serious problems with the lazy I/O:

- The order of the side effects is tied to the order of the lazy evaluation. Because the order of lazy evaluation is not explicit, the order of effects also isn't. The sequence of side-effects can become hard to predict.

- It can be difficult to reason about the space requirements of a lazy program. For example, in the previous example, when we use `print $ words contents`, all the file contents will be held in memory at once. We could have used `hGetContents h >>= return . lines` to print the lines incrementally, thereby using only a constant space. This shows that the space requirements are contextual to how the stream is used.

- Poor resource management and lack of explicit order of effects can make it difficult to know when to clean up resources. Since the demand drives lazy evaluation, which drives effects, we inherently have little opportunity to "intercept" evaluation for resource management purposes. Also, resource management is made more difficult by the possibility of errors. We will see more about this in the next section *Resource management with Bracket*.

Despite this, the lazy I/O remains an attractive option in simple situations, where the space requirements and order of execution are sufficiently predictable and where resource management is easy enough. However, when there is strong demand for precise resource management or predictable space usage, lazy I/O is not an option; for example, for writing networking code, handling many files, or handling many HTTP requests in a web server, and so on.

> *"Extensive experience in Haskell has, however, exposed severe drawbacks of lazy evaluation, which are especially grievous for stream processing of large amounts of data.*
>
> *Lazy evaluation is fundamentally incompatible with computational effects, can cause fatal memory leaks, and greatly inhibits modular reasoning, especially about termination and space consumption.*
>
> *Seemingly innocuous and justified changes to the code or code compositions may lead to divergence, or explosion in memory consumption."*
>
> *Lazy v. Yield: Incremental, Linear Pretty-printing - Kiselyov et al*

Before we look at an established solution to the problem of lazy I/O (Iteratee I/O), we'll briefly explore the basic lazy I/O approach to resource management in the face of exceptions.

Resource management with bracket

So far, we have been explicitly opening and closing files. This is what we call explicit resource management:

```
main = do
  h <- (openFile "jabberwocky.txt" ReadMode)
  useResource h
  hClose h
where
    useResource h'
        = (stream h') >>= mapM_ putStrLn
    stream h'
        = hGetContents h' >>= return . lines
```

Let's look at some higher level abstractions to capture this pattern: open resource, use it, in some way clean up resource. The crudest solution is to just ignore the problem and rely on the garbage collector for the cleanup:

```
main = do
  contents <- readFile "jabberwocky.txt"
  mapM_ putStrLn (lines contents)
```

The `readFile` function encapsulates the file handle, which is then garbage collected when the `contents` function is garbage collected or when it has been entirely consumed. This is very poor resource management!

It would be more idiomatic to use the wrapper function `withFile`:

```
main = do
  withFile "jabberwocky.txt" ReadMode enumerateLines
  where
    enumerateLines h = lines' h >>= mapM_ putStrLn
    lines' h' = hGetContents h' >>= return . lines
```

The `withFile` function cleanly decouples the producer from the consumer and gives better control over resource management. The file will be closed in case of completion or error because `withFile` makes use of the `bracket` function:

```
bracket
  (openFile "filename" ReadMode) -- acquire resource
  hClose                         -- release resource
  (\h -> "do some work")
```

where

```
bracket ::  IO a                -- before action
        -> (a -> IO b)  -- after action
      -> (a -> IO c)  -- do action
      -> IO c              -- result
```

The `bracket` function relies on higher order functions to express a specific kind of wrapper pattern: "acquire and release". For more information, visit `https://wiki.haskell.org/Bracket_pattern`.

The `finally` function is a special form of bracket:

```
finally :: IO a       -- some action
          -> IO b    -- final action: runs afterwards
          -> IO b  -- result
```

The bracket family of functions helps us clean up resources more reliably, but by no means definitively solves the problem of closing resources in a timely manner. If we need more precise resource management than this (and more predictable space requirements and ordering of effects), then we must use a more sophisticated pattern for stream programming called **Iteratee I/O**.

Iteratee I/O

Why Functional Programming Matters [REF 1990, *John Hughes*] has been described as the manifesto for lazy programming, written at a time when enthusiasm for this style was running very high. Less than 20 years later, Oleg Kiselyov published the obituary of Lazy I/O in a series of writings; for more information, visit `http://okmij.org/ftp/Streams.html`.

In the late 2000s, Kiselyov championed a new way of doing I/O that combines the best of Handle-based I/O (precise control over resources and predictable space requirements) with the best of lazy I/O (decoupling of producers and consumers, high level of abstraction).

Let's get to the root of this style of programming by rephrasing the example we saw earlier. Recall the `Chunk` data type and the `parseChunk` function:

```
data Chunk = Chunk {chunk :: String}
                  | LineEnd {chunk :: String,
                                    remainder :: String}

parseChunk :: ByteString -> Chunk
parseChunk chunk
 = if rightS == B8.pack ""
      then Chunk   (toS leftS)
      else LineEnd (toS leftS) ((toS . B8.tail) rightS)
  where
     (leftS, rightS) = B8.break (== '\n') chunk

-- this time we extract this helper function to the top level:
toS = map (chr . fromEnum) . B.unpack
```

We will iterate through the chunks in a file and accumulate them into lines. Each time a new line is accumulated, we will do some I/O with that line.

Let's model the iteration step as a function that processes a file chunk and returns an `IterResult` value:

```
-- iterF :: String -> IterResult
```

 This example is available at `http://www.scs.stanford.edu/11au-cs240h/notes/iteratee.html` and `https://themonadreader.files.wordpress.com/2010/05/issue16.pdf`; look for *Iteratee: teaching an old fold new tricks.*

Because we want to refer to this function signature in `IterResult`, we capture it as a data type:

```
newtype Iter
    = Iter {runIter :: B8.ByteString -> IterResult}
```

This type represents the concept of an "iteratee". The `IterResult` function can indicate two possibilities, `HaveChunk` and `NeedChunk`:

```
data IterResult
    = HaveLine {line :: String, residual :: String}
    -- a line has been accumulated (with possible residual)
    | NeedChunk Iter
    -- need more chunk data

instance Show IterResult where
    show (HaveLine l r) = "HaveLine " ++ l ++ "|" ++ r
    show (NeedChunk _) = "NeedChunk"
```

The `HaveLine` function simply returns some data fields, but the `NeedChunk` function is not so simple:

```
NeedChunk (String -> IterResult)
```

Instead of returning a value, the `NeedChunk` function returns an `Iter` function, that is, another step function. This is at the heart of Iteratee I/O: a step function can return another step function. The easiest way to make sense of this is to continue a bit further down this road.

Iteratee

Next, we'll write an `Iter` function:

```
chunkIter :: Iter
chunkIter = Iter (go "")
    where
        go :: String -> B8.ByteString -> IterResult
        go acc chunk =
    case (parseChunk chunk) of
        (Chunk chunk')
            -> NeedChunk (Iter (go (acc ++ chunk')))
        (LineEnd chunk' residual')
            -> HaveLine (acc ++ chunk') residual'
```

Note that when we curry go with the `acc` parameter in `go acc`, we get the `Iter` type signature:

```
(B8.ByteString -> IterResult)
```

Let's run `chunkIter` function on a chunk of file. To start with the simple case, we make the chunk size large enough so that we can expect our `IterResult` function to return a `HaveLine`, which contains a line with a residual chunk:

```
main = do
  h <- openFile "jabberwocky.txt" ReadMode
  chunk1 <- B.hGet h 50
  print $ runIter chunkIter chunk1
-- HaveLine "'Twas brillig, and the slithy toves"
--                    "Did gy"
```

With a chunk size smaller than the first file line, we get an `IterResult` of the `NeedChunk` function instead:

```
main = do
    h <- openFile "jabberwocky.txt" ReadMode
    chunk1 <- B.hGet h 25
    print $ runIter chunkIter chunk1
    -- NeedChunk
```

The first run of the `chunkIter` function returns `NeedChunk`. This is the iteratee's way of communicating to us (the caller) that we need to perform another iteration. Instead of asking us to simply call the `chunkIter` function again (which has no knowledge of the chunks of data accumulated so far), the iteratee provides us with the next iteratee to run (which has embedded knowledge of accumulated chunks):

```
main = do
    h <- openFile "jabberwocky.txt" ReadMode

    chunk1 <- B.hGet h 25
    let (NeedChunk iter1) = runIter chunkIter chunk1

    chunk2 <- B.hGet h 25
    let (HaveLine line residual) = runIter iter1 chunk2
    putStrLn line
```

Enumerator

Let's write a function that will loop through the chunks of a file and accumulate lines. An enumerator feeds data to an iteratee to get a result:

```
enumerateFile path initIter =
  withFile path ReadMode $ \h ->
    let
    go iter = do
      isEOF <- hIsEOF h
      if isEOF
        then return (HaveLine "End Of File" "")
        -- we're cheating
        else do
          chunk <- B.hGet h 8
          check $ runIter iter chunk
check (NeedChunk iterNext)
= go iterNext
check (HaveLine line residual)
= do
    putStrLn line
    check $ runIter initIter
                            (B8.pack residual)
    in go initIter
```

The enumerator takes an initial iteratee, runs it, and checks the result. If the iteratee needs another chunk, the enumerator runs the step function embedded in the iteratee result. If the iteratee signals that it has accumulated a line, the enumerator processes the line and passes the residual to the original iteratee (since the HaveLine function case does not return a step function to run in our design).

We can use our enumerator like this:

```
main = do enumerateFile "jabberwocky.txt" chunkIter
```

The enumerator has a type:

```
enumerateFile :: FilePath -> Iter -> IO IterResult
```

If we apply the first argument to enumerateFile, we get the type:

```
(enumerateFile file) :: Iter -> IO IterResult
```

We can concretize this type as follows:

```
type Enumerator = Iter -> IO IterResult
```

This shows that an enumerator takes an iteratee, performs the iteration with possible side effects, and then returns the iteratee result:

```
enumerateFile :: FilePath -> Enumerator
```

Enumerators work like folds, in the sense that they apply a function (with an accumulator) over an input stream element by element.

Generalized iteratees, enumerators, and enumeratees

Our code is flawed, and a very long way from what we would need for a robust and composable Iteratee I/O.

For instance, with the preceding code, the last line of the file is lost. This is easily solved by passing a richer input type to the iteratee. Instead of the chunk `ByteString`, we can have the following code:

```
data IterInput = Chunk' String | EndOfFile
```

At the end of the file, our enumerator can pass in the `EndOfFile` function to the iteratee and ask it to return the last line as the `HaveLine` function.

To address the critical issue of dealing with failure, the iteratee can also signal failure to the enumerator by extending the `IterResult` function:

```
data IterResult
  = HaveLine {line :: String, residual :: String}
  | NeedChunk Iter
  | Failure {errMsg :: String}
```

In our example, the iteratee and enumerator types were hardcoded to the types we needed, but these can easily be generalized by parameterizing our types.

We also want to be able to compose iteratees and enumerators; this can be achieved by making them implement the monad type class.

Another major omission is that, in practice, we also need the enumeratee's abstractions that enable transforming the output of an enumerator or iteratee and feeding that into another iteratee. This allows for very flexible processing pipelines:

```
--  ITERATEE   -> ENUMERATOR
--  ITERATEE1 -> ITERATEE2  -> ENUMERATOR1 -> ENUMERATOR2 ...
--  ITERATEE   -> ENUMERATEE -> ENUMERATOR
--  ... -> ENUMERATEE1 -> ENUMERATEE2 -> ENUMERATEE3 ...
```

Iteratees produce data, enumeratees serve as pipeline transformers of data, and enumerators consume data and drive the whole pipeline process.

While the enumerator drives, the iteratee can also influence evaluation by signaling when it is done processing, when it needs more data, when it can yield a result, or when it has encountered a failure. The producer collaborates with the consumer.

With this flexibility, resource management in the face of exceptions becomes much more tractable than with lazy I/O, and it can be abstracted at a much higher level than with Handle-based I/O.

> *"Iteratee I/O is a style of incremental input processing with precise resource control. The style encourages building input processors from a user-extensible set of primitives by chaining, layering, pairing and other modes of compositions.*
>
> *The style is especially suitable for processing of communication streams, large amounts of data, and data that has undergone several levels of encoding such as pickling, compression, chunking, framing.*
>
> *It has been used for programming high-performance (HTTP) servers and web frameworks, in computational linguistics and financial trading."*
>
> *- Iteratees, Kiselyov*

The iteratee I/O libraries

Iteratee libraries differ markedly in how they model types for iteratees, enumerators, and enumeratees. Even the meaning and necessity OF enumeratees varies between libraries. Libraries also differ in the type of signals the iteratee can send back to the consumer (that is, the kind of influence the iteratee can have on an evaluation).

There are several libraries to choose from; the choice is made simpler by understanding the chronology and historical development:

- Oleg Kiselyov laid down the groundwork in his early IterateeM/CPS libraries.

- John Lato, together with Oleg Kiselyov, clarified some key concepts in the "iteratee" library (`https://hackage.haskell.org/package/iteratee`)

- The next wave brought John Millikin's popular "enumerator" library (`https://hackage.haskell.org/package/enumerator`) and "iterIO", which is to be commended for its simplified mental model (great for learning about Iteratee I/O). [REF: `https://hackage.haskell.org/package/iterIO`]

- As of this writing, the most "modern" Iteratee I/O libraries are "pipes" (Gabriel Gonzales), which focus on preserving equational reasoning (`https://hackage.haskell.org/package/pipes`), and "conduit" (Michael Snoyman), which focusses on deterministic resource management, arguably an issue that is more important to resolve (`https://hackage.haskell.org/package/conduit`)

Comparing the three styles of I/O

Let's compare the 3 styles of I/O we have encountered from a few perspectives:

	Handle-based I/O	Lazy I/O	Iteratee I/O
Processing	Strict and incremental	Lazy	Strict and incremental
What drives evaluation?	Looping code	Stream consumer	Enumerator + Iterator
Level of abstraction	Low level	High level	High level
Resource management, dealing with Exceptions	Precise control	No precise control	Precise control

Summary

We explored three styles of doing I/O in Haskell: imperative I/O, lazy I/O, and Iteratee I/O. While imperative I/O gave us fine control over resource management and space requirements, it had to be written at a low level of abstraction.

In contrast to this, lazy I/O stood out as elegant and written at a high level of abstraction, giving us great decoupling between producers and consumers. Then we explored the high price we pay for relinquishing control, compared to when side effects occur.

Finally, we found that Iteratee I/O gives us the best of both worlds, fine control over resource management and usage, and is also written at a very high level of abstraction.

While looking at the three styles, we paid close attention to who was driving evaluation. We looked at the means of decoupling producers and consumers of data and also the extent to which producers and consumers can communicate in the midst of data processing.

In the next chapter, we will focus on patterns for composition with a focus on the fundamental types: functor, applicative functor, arrow, and monad. There is much more to compose in Haskell than functions!

3
Patterns of Composition

Function composition is a fundamental part of functional programming.
This chapter is concerned with exploring the composition characteristics of the fundamental type-classes: functor, applicative functor, arrow, and monad. We will see that functor embeds into applicative functor, which embeds into arrow, which embeds into monad. After exploring monad composition, we'll also look into monad transformers (a technique for composing different types of monads). As we move through the successive types, from the most general (functor) to the most powerful (monad), we will see how they differ in the ways they can be composed.

Note that this chapter does not have the last word on composition patterns, for example, in the following chapter we will explore another pattern for composition: functional lenses.

- Functor
- Applicative functor
- Monad
- Monad transformers
- Arrows

Functor

The Functor type-class gives us a way to generalize function application to arbitrary types. Let's first look at regular function application. Suppose we defined a function of primitive types:

```
f :: Num a => a -> a
f = (^2)
```

We can apply it directly to the types it was intended for:

```
f 5
f 5.0  -- etc          .
```

To apply the f variable to a richer type, we need to make that type an instance of the Functor class and use the fmap function:

```
--  fmap function    Functor
fmap       f                (Just 5)
fmap       (f . read) getLine
```

The Functor class defines fmap:

```
class Functor f where
  fmap :: (a -> b) -> f a -> f b
```

Let's create our own Maybe' type and make it an instance of Functor:

```
data Maybe' a = Just' a | Nothing'
  deriving (Show)

instance Functor Maybe' where
  fmap _ Nothing' =   Nothing'
  fmap f (Just' x) = Just' (f x)
```

By making Maybe' a Functor class, we are describing how single-parameter functions may be applied to our type, assuming the function types align with our Functor class, for example:

```
-- we can do this
fmap f (Just' 7)
fmap show (Just' 7)

-- but still not this
fmap f (Just' "7")
```

The fmap acts as a lifting function, taking our function up into the realm of the Functor. The fmap also lifts function composition to the level of functors. This is described by the **functor laws**:

```
-- law of composition
fmap (f . g)  ==  fmap f . fmap g

-- e.g.
fmap (f . read) getLine  -- is the same as
```

```
(fmap f) . (fmap read) $ getLine

-- identity law
fmap id  ==  id
-- e.g.
fmap id (Just 1) = id (Just 1)
```

The `fmap` is to `Functor` what `map` is to the `List` type, as shown in the following code:

```
ns = map (^2) [1, 2, 3, 5, 7]
-- 'map' lifts f to operate on List type
```

We can also write the preceding code as follows:

```
ns' = fmap (^2) [1, 2, 3, 5, 7]
```

because List is a `Functor`:

```
instance Functor List where
   fmap = map
```

The `Functor` class abstracts the idea of function application to a single argument. This class give us a way of combining functions with types by lifting a function from one level of abstraction to another.

Applicative functor

Because `Maybe` is a `Functor`, we can lift the `(+2)` function so that it can be applied directly to a Maybe value (Just or Nothing):

```
fmap (+2) (Just 3)
```

However, fmap does not enable us to apply a function to multiple `Functor` values:

```
fmap (+) (Just 2) (Just 3)
```

For that, we need the `Applicative Functor` class, which enables us to raise a function to act on multiple `Functor` values:

```
-- Applicative inherits from Functor
class (Functor f) => Applicative f where
   pure  :: a -> f a
   (<*>) :: f (a -> b) -> f a -> f b
```

The `pure` function lifts a value into the `Functor` class, and the `<*>` operator generalizes function application to the `Functor` class (hence `Applicative Functor` class). Let's see how this works by making `Maybe'` an instance of `Applicative`:

```
import Control.Applicative

data Maybe' a = Just' a | Nothing'
  deriving (Show)

-- we still need the Functor instance
instance Functor Maybe' where
  fmap _ Nothing' =   Nothing'
  fmap f (Just' x) = Just' (f x)

instance Applicative Maybe' where
  pure f = Just' f
  Nothing'  <*> _          = Nothing'
  _         <*> Nothing'   = Nothing'
  (Just' f) <*> (Just' x)  = Just' (f x)
```

The `Applicative` class operators `pure` and `<*>` give us a way of lifting pure functions up to the `Applicative Functor` level, for example:

```
pure  (,) <*> Just' 2 <*> Just' 3
-- evaluates as
Just' (,) <*> Just' 2 <*> Just' 3
Just' ((,) 2)          <*> Just' 3
Just' ((,) 2 3)
Just' (2,3)
```

The currying of the lower-level function `(,)` leads to currying on the `Applicative` level. Similarly, function composition on the lower level is preserved at the `Applicative` level:

```
Just' (.) <*> Just' (+2) <*> Just' (+3) <*> Just' 1
-- evaluates as
Just' ((.) (+2))          <*> Just' (+3) <*> Just' 1
Just' ((.) (+2) (+3))                    <*> Just' 1
Just' (((.) (+2) (+3)) 1)
Just' 6
```

In fact, this expresses the law of composition of `Applicative Functor`:

```
pure (.) <*> u <*> v <*> w =  u <*> (v <*> w)


-- e.g.
Just' (.) <*> Just' (+2) <*> Just' (+3) <*> Just' 1
-- is the same as
Just' (+2) <*> (Just' (+3) <*> Just' 1)
```

When we combine two `Applicative`'s with the `<*>` operator, we always get another `Applicative`. In other words, `Applicative`'s are "closed under composition".

Since `Applicative` is a `Functor`, we can also use the `fmap` function:

```
pure  (,) <*> Just' 2 <*> Just' 3
-- same as
(fmap (,) (Just 2))   <*> Just' 3
((,) <$> (Just 2))    <*> Just' 3
-- same as
(,)   <$> (Just 2)    <*> Just' 3
```

Here `<$>` is an infix synonym for `fmap`. Used in this way, `Applicative` provides us with a multiparameter `fmap` function. The `<*>` operator is more generic and uniformly lifts function application to the `Functor` level:

```
fmap, (<$>) ::  (a -> b) -> f a -> f b
(<*>)       :: f (a -> b) -> f a -> f b
```

Monad

The `Monad` class inherits from the `Applicative` class (only from GHC 7.10 onward; see the *Monad as applicative* section for more on this):

```
class (Applicative m) => Monad m where
  return :: a -> m a
  (>>=) :: m a -> (a -> m b) -> m b
```

The `return` function looks just like the `pure` function of the `Applicative` class (it wraps a value in a `Monad` class).

The `bind` operator (`>>=`) combines a `Monad` (`m a`) with a function (`a -> m b`), which we'll call a **monadic function**. The monadic function acts on type `a` of the first monad and returns a new monad of type (`m b`).

Let's make our `Maybe'` type a `Monad` class:

```
import Control.Monad
import Control.Applicative

data Maybe' a = Just' a | Nothing'
  deriving (Show)

instance Functor Maybe' where
-- ...

instance Applicative Maybe' where
-- ...

instance Monad Maybe' where
  return x = Just' x
  Nothing'  >>= _   = Nothing'
  (Just' x) >>= f   = (f x)
```

The `bind` operator for `Maybe'` says:

- Given `Nothing'`, ignore the monadic function and simply return `Nothing'`
- Given `Just' x`, apply the monadic function to the value `x` ; this will return a `Maybe'` monad of possibly different type

Let's use bind with our `Maybe'` Monad:

```
-- monad >>= monadic function
Just' 10 >>= \x -> Just' (show x)

-- evaluates as
(\x -> Just' (show x)) 10
Just' "10"
```

Here is another example:

```
Nothing' >>= \x -> Just' (show x)
-- evaluates as
Nothing'
```

At first glance, `bind` simply allows us to combine monads with monadic functions. However, as every Haskell journey man and -woman has found, things are not quite that simple. We will talk more about `bind` soon, but let's first explore how monad relates to `Functor` and `Applicative`.

Monad as functor

We can change Monad to the Functor class by using the `liftM` function, the monadic version of `fmap`:

```
instance Functor Monad where
    fmap = liftM
```

where

```
liftM :: Monad   m => (a -> b) -> m a -> m b
fmap  :: Functor m => (a -> b) -> m a -> m b
```

Now we have three lines of expressing Functor-like behavior:

```
main = do
  -- Levels of Functors
  print $ fmap (*2) (Just' 10)       -- FUNCTOR
  print $ pure (*2) <*> (Just' 10) -- APPLICATIVE
  print $ liftM (*2) (Just' 10)      -- MONAD
```

The monad `liftM` function is based on the `bind` operator:

```
liftM f m = m >>= return . f

-- in other words
liftM f m = do
    val <- m         -- extract value
    return (f val) -- wrap result in Monad

-- e.g.
liftM (*2) (Just' 10)
= (Just' 10) >>= return . (*2)
= return . (*2) 10
= return 20
= Just' 20
```

For example, extract the Monad value, pass it to f, and lift the result into the Monad value with `return`.

Monad as applicative

Applicatives can lift functions of many arguments. Monads can also do so, albeit less elegantly, with the `liftM` functions (`liftM`, `liftM2`, `liftM3`, ...):

```
main = do
  print $ (<$>)  (*) (Just' 10) <*> (Just' 20) -- APPLICATIVE
  print $ liftM2 (*) (Just' 10)    (Just' 20) -- MONAD
```

Any monad is also an applicative functor:

```
-- ap_ defines <*> for Monads
ap_ mf mx = do
  f <- mf       -- extract function
  x <- mx       -- extract val
  return (f x)
```

that is, extract the function from the first `Monad` and the value from the second, and do the function application (this method already exists as `Control.Monad.ap`).

Now we can write monadic code in applicative style:

```
(Just' (*)) 'ap_' (Just' 10) 'ap_' (Just' 20)
```

We can easily make `Monad` an instance of `Applicative`:

```
instance Applicative Monad where
  pure = return
  (<*>) = ap
```

The "applicative pattern" was recognized and extracted as `Applicative` only in 2008, more than ten years after `Monad` became an established part of Haskell (and 20 years after Eugenio Moggi's `Monad` paper).

This is why we have several ways of doing the same thing, as shown in the following table:

Functor (1990)	Applicative (2008)	Monad (1990s)
fmap	pure, <*>	liftM
	<*>	ap
	pure	return

These discrepancies have been resolved by the "Functor-Applicative-Monad Proposal", implemented in base 4.8.0.0 used by GHC 7.10 and above (https://wiki.haskell.org/Functor-Applicative-Monad_Proposal).

Sequencing actions with monad and applicative

Monad can sequence actions as follows:

```
action s = do putStrLn s; return s
main = do
  let actions = map action ["parts", "are", "disconnected"]
  sequence' actions
  return ()
```

Here `sequence'` performs the actions one after the other:

```
sequence' [] = return []
sequence' (x:xs) = do
  x'  <- x                -- action performed
  xs' <- sequence' xs
  return (x':xs')
```

But we can also sequence actions with `Applicative`:

```
sequenceA [] = pure []
sequenceA (x:xs) = (:) <$> x <*> (sequenceA xs)
-- sequenceA actions
```

(In fact, a part of the Functor-Applicative-Monad Proposal mentioned earlier is to change the prelude's `sequence` function to require applicative instead of monad.)

An `Applicative` class can sequence actions that happen in isolation, that is, actions that don't depend on the results of previous actions. But when actions in a sequence need to communicate results to subsequent actions, `Applicative` becomes insufficient and we need `Monad`.

Monads and the bind chain

The magic of `Monad` is that we can build a chain of actions such that an action can communicate with subsequent actions. This is not so with `Applicative`; for example, you cannot express this directly with `Applicative`:

```
main = do
  line <- getLine                -- ACTION 1
  putStrLn $ "You said " ++ line -- ACTION 2
            -- uses result of ACTION 1
```

The `bind` operator (`>>=`) lets us bind outputs to inputs, whereas with `Applicative` (`<*>`), each action is performed in isolation (there is no "contact point" between two actions).

An important implication of this is that, with monads, we can have a dynamic sequence of actions, where the outcome of one action can affect which subsequent actions are performed, for example:

```
main = mainLoop
mainLoop = do
  line <- getLine                    -- ACTION 1
  if line == "stop"
    then putStrLn "Bye"              -- ACTION 2b
    else do
      putStrLn $ "You said " ++ line -- ACTION 2c
      mainLoop
```

This cannot be expressed with `Applicative`, where each action in the sequence is destined to be performed. It is worth saying, however, that `Applicative` does allow for limited communication between actions, for example:

```
(+) <$> Nothing <*> Just 10 <*> Just 20
```

Here the first instance of `Nothing` will prevent all subsequent actions from being performed. We have essentially baked the communication between actions directly into the `Maybe Applicative` type instance.

We saw earlier, while working with the **Reader Monad**, that we can use the bind chain to provide a "shared context" between actions in a sequence.

The shared context can be used as a place to do "out of band" processing, that is, processing that is made explicit in the bind chain but remains implicit from the perspective of the chain of monadic actions. As an example, consider a sequence of actions occurring in the context of a Reader Monad: the reader state is out of band, that is, "independent" of the monadic pipeline.

This explains why we can use monads to approximate imperative programming (where out of band processing is so prevalent).

In this sense, we can view the bind chain as a more sophisticated version of an accumulator argument in a tail recursive function (the accumulator allows for a shared context in a nested chain of recursive function calls).

Monads bind in a way that includes an "accumulator". This contrasts with applicatives, which have no accumulator and hence no communication between arguments.

The `bind` operator composes `Monad` values with monadic functions (functions that return a value embedded in a `Monad` class). With the `Functor` and `applicative` classes, the functions were ignorant of the types they were being lifted to. When chaining a `Monad` class with a monadic function, the function is entwined with the `Monad` class it is being chained with.

Also, it is worth nothing that monads are not "closed under composition", as is the case with applicatives, because monads generally don't compose into monads.

Composing with monads

Let's summarize the ways in which we can compose with monads:

- We can compose pure functions with monads:

  ```
  liftM* f m -- returns another Monad m
  ```

- We can compose monadic functions with monads:

  ```
  m >>= fM >>= gM >>= hM
  ```

- We can compose monadic functions with each other:

  ```
  gM <=< fM
  ```

  where (`<=<`) is syntactic sugar for

  ```
  gM <=< fM = \x -> (fM x) >>= gM
  ```

The key composition is binding (`>>=`) the monad with a monadic function, but beyond that monads don't compose as well as applicatives.

Monad transformers

In this section, we look at combining different types of monads into more powerful combinations. We can do this by creating "monad stacks". Let's start with a simple Reader Monad and then stack some other monads on top of it.

We'll use a Reader Monad to hold some configuration data for an application:

```
data Config = Config {discountRate :: Float,
                                 currencySym  :: String}
    appCfg = (Config 10 "R")
```

The `discount` function takes a `Float` value and returns another `Float` value, but in the context of a `Reader Config` variable:

```
discount :: Float -> Reader Config Float
discount amt = do
    discountRate' <- asks discountRate
    return (amt * (1 - discountRate' / 100))
```

From within the function, we can ask for the configuration data.

Now we can use the `runReader` function with specific configuration data:

```
import Control.Monad.Identity
import Control.Monad.Reader
import Control.Monad.Writer

main = do
  print $ runReader (discount 100) appCfg
```

Let's add a `display` function in the context of `Reader Config`:

```
display :: Float -> Reader Config String
display amt = do
  currencySym' <- asks currencySym
  return (currencySym' ++ " " ++ (show amt))

main = do
  putStrLn $ runReader doDoubleDiscount appCfg
 where doDoubleDiscount
    = (discount 100 >>=
        discount >>=
        display)
```

We have a chain of monadic functions executing in the context of `Reader Config`. The result is fed into the `putStrLn` function, which executes in the IO Monad. To add logging capability to our "Reader functions", we can stack our `Reader` type on top of a `Writer` Monad.

The `ReaderT` type is a Reader Monad that also takes an inner monad, in this case `Writer String`. The rest of the function is still valid as is, but now we can access the writer's `tell` function from inside our functions:

```
discountWR :: Float -> ReaderT Config (Writer String) Float
discountWR amt = do
  discountRate' <- asks discountRate
  let discounted = amt * (1 - discountRate' / 100)
 tell $ "-Discount " ++ (show amt) ++ " = "
```

```
          ++ (show discounted)
    return discounted

displayWR :: Float -> ReaderT Config (Writer String) String
displayWR amt = do
  currencySym' <- asks currencySym
  tell " > Displaying..."
  return (currencySym' ++ " " ++ (show amt))

main = do
  print $ runWriter (runReaderT doDoubleDiscount
appCfg)
where doDoubleDiscount
= (discountWR 100 >>=
    discountWR >>=
    displayWR)
```

Since we have a `Reader` wrapped around a Writer Monad, we need to first use `runReaderT` (which unwraps the Reader Monad and gives us the result):

```
    runReaderT someApp appCfg
```

The `Reader` function's result is the inner Writer Monad, which needs to be unwrapped to get the result:

```
    runWriter (runReaderT someApp appCfg)
```

In this way, `Monad` stacks are unwrapped in the opposite order to that of the "wrapping".

But this is becoming messy:

```
discountWR :: Float -> ReaderT Config (Writer String) Float
displayWR  :: Float -> ReaderT Config (Writer String) String
runWriter (runReaderT someApp appCfg)
```

We can simplify things with the following line of code:

```
    type App = ReaderT Config (Writer String)
```

Note the currying of type parameters. We omit the final `ReaderT` type because we want to vary it:

```
    discountWR :: Float -> App Float
    displayWR  :: Float -> App String
```

Now we can also define the doApp function:

```
doApp :: App a -> (a, String)
doApp app = runWriter (runReaderT app appCfg)
```

```
main = do
  print $ doApp doDoubleDiscount
  where doDoubleDiscount = (discountWR 100 >>=
          discountWR >>=
          displayWR)
```

More idiomatically, we can use the newtype method:

```
newtype App a = App {runApp :: ReaderT Config (Writer String) a}
  deriving (Monad, MonadReader Config, MonadWriter String)
```

The deriving clause requires the following LANGUAGE pragma (at the top of the file):

```
{-# LANGUAGE GeneralizedNewtypeDeriving #-}
```

We are making App an instance of all these type-classes. MonadReader is the class that implements the Reader functionality: both Reader and ReaderT implement MonadReader. Similarly, Writer and WriterT type implement MonadWriter. This allows for code sharing between the monads and their transformer versions.

The effect of this is to "flatten" our nested stack of monads, making all lower-level functions available on the top level (for example. we can call tell and asks as if they are defined on the App monad level of the stack).

Without this convenience measure, we would have to lift the nested Monad function once for each level of the stack, until we reached the monad with the function we are after. A "flattened" nest of monads lets us avoid excessive lifting through the stack's layers.

On the downside, this elegant way of simplifying the practicalities of working with stacks is precisely what makes it more tedious to define your own monad transformers. Having said that, there are more recent packages that attempt to solve this problem. See **extensible-effects** (https://hackage.haskell.org/package/extensible-effects) and **layers** (https://hackage.haskell.org/package/layers) for more information on this.

Returning to our example, with the newtype App variable, our function type signatures remain the same:

```
discountWR :: Float -> App Float
displayWR :: Float -> App String
```

Also, by using `newtype` in this way, we can limit what we export from our code and thereby obfuscate the details of how the stack is constructed. However, in `doApp` we now need to use the `runApp` app:

```
doApp :: App a -> (a, String)
doApp app = runWriter (runReaderT (runApp app) appCfg)

main = do
  print $ doApp doDoubleDiscount
  where doDoubleDiscount = (discountWR 100 >>=
          discountWR >>=
          displayWR)
```

IO in monad stacks

We can also add the IO Monad to our `Monad` stack, although we will see that IO is a special case:

```
newtype AppIO a
  = AppIO {runAppIO :: ReaderT Config (WriterT String IO) a}
  deriving (Monad, MonadReader Config,
              MonadWriter String, MonadIO)
```

Instead of `Writer String`, we use `WriterT String IO`, that is, a Writer Monad wrapping IO. Also, we've added a `derive` clause for `MonadIO`, which is to IO what `MonadReader` is to the `Reader` Monad. `MonadIO` adds IO access to all `Monad` in the stack.

```
discountWRIO :: Float -> AppIO Float
displayWRIO  :: Float -> AppIO String
```

The `doAppIO` function returns the previous result but is now wrapped in an IO action:

```
doAppIO :: AppIO a -> IO (a, String)
-- use runWriterT to extract the writer result
doAppIO app = runWriterT (runReaderT (runAppIO app) appCfg)
```

Now we can do IO operation in our `Monad` stack functions. To expose the IO Monad, we need to use the `liftIO` function:

```
discountWRIO amt = do
  liftIO $ putStrLn "We're doing IO!"
  discountRate' <- asks discountRate
  let discounted = amt * (1 - discountRate' / 100)
  tell $ " > Discounting " ++ (show amt)
          ++ " = " ++ (show discounted)
```

```
    return discounted

displayWRIO amt = do
  liftIO $ putStrLn "More IO!"
  currencySym' <- asks currencySym
  tell " > Displaying..."
  return (currencySym' ++ " " ++ (show amt))

main = do
  print <$> doAppIO doDoubleDiscount
where doDoubleDiscount
  = (discountWRIO 100 >>=
       discountWRIO >>=
       displayWRIO)
```

Sequence of stack composition

A Monad stack implies a sequence of composition of Monads. In the previous example, we had a Reader Monad wrapping a Writer Monad. We could just as well have swapped the position of `Writer` and `Reader` in our stack:

```
newtype AppIO a
  = AppIO {runAppIO :: WriterT String (ReaderT Config IO) a}
deriving (Monad, MonadReader Config,
            MonadWriter String, MonadIO)
```

This means we would swap around the order of unwrapping the stack:

```
doAppIO :: AppIO a -> IO (a, String)
doAppIO app = runReaderT (runWriterT (runAppIO app)) appCfg
```

The order of `Reader` and `Writer` is inconsequential, since these monads are unaffected by each other in a `Monad` stack. This is not true for all combinations of monads, however. If one monad relies on the work of a previous monad being done, the order indeed matters (as with the composition of functions).

Moreover, `IO` is a special case and must remain at the bottom of the stack.

Arrows

Let's find our way towards arrows from the perspective of monads. Look at the following IO code:

```
import System.IO
main = do
liftM (length . words)
        (readFile "jabberwocky.txt" ) >>=
        print
   -- regular functions: length, words
   -- Monadic functions: readFile, print
```

We use `liftM` to lift the the composed function `length . words` into the monadic function `readFile` and then feed the result to another monadic function `print`.

We can compose the regular functions with `(.)`, but we know well that we cannot do the following:

```
print . length . words . readFile "jabberwocky.txt"
-- INVALID - types don't align
```

Let's make the preceding code possible!

> The following code is based on a combination of *Programming with Arrows* by *John Hughes*, and a blog post by John Wiegley at `http://www.newartisans.com/2012/10/arrows-are-simpler-than-they-appear/`.

The crux of our approach will be to create a "meta type" to represent monadic IO functions and then define composition for that type. Our meta type for monadic IO functions is as follows:

```
data IOF a b = IOF {runIOF :: a -> IO b}
```

`IOF` wraps a function `(a -> IO b)` and places the input and output types `a` and `b` on an equal footing, while also hiding the IO Monad. Next, we define our own composition operator:

```
(<<<<) :: IOF a b -> IOF c a -> IOF c b
(IOF f) <<<< (IOF g) = IOF $ f <=< g
```

This function takes two `IO` functions, composes them, and wraps the resulting `IO` function.

Finally we need a function to lift a `Monadic IO` function into an `IOF` variable:

```
lift' :: (a -> b) -> IOF a b
lift' f = IOF $ return . f -- uses IO Monad's return
```

Now we can compose regular and `IO` functions:

```
main = do
  let f =  IOF print <<<< lift' length <<<<
  lift' words <<<< IOF readFile
  runIOF f "jabberwocky.txt"
  return ()
```

By doing this, we have started to reinvent `Arrows`. Let's do the same as we just did, but this time using the arrow type-class.

Implementing an arrow

To write our first arrow, we'll need the IOF type from the previous example. However, this time we use a better name for it: `IOArrow`. This time we call our meta type for `IO` functions `IOArrow`:

```
data IOArrow a b = IOArrow {runIOArrow :: a -> IO b}
```

To make `IOArrow` a true `Arrow`, we need to implement `Category` and `Arrow`. `Category` describes function composition `(.)`.

```
instance Category IOArrow where
  id = IOArrow return
  -- (.) = (<<<<)
  IOArrow f . IOArrow g = IOArrow $ f <=< g

instance Arrow IOArrow where
  -- arr = lift'
  arr f = IOArrow $ return . f
  first (IOArrow f) = IOArrow $ \(a, c) -> do
    x <- f a
    return (x, c)
```

The `arr` function is just `lift'` from earlier (using `IOArrow` instead of `IOF`). Recall that return lifts a value into a Monad and pure lifts a value into an Applicative. The `arr` function is to `Arrow` what `return` and `pure` are to `Monad` and `Applicative` classes, respectively.

Before we delve into the meaning of the `first` function and the other key `Arrow` operators, let's use our `IOArrow` function:

```
import Prelude hiding ((.), id)
import Control.Category
import Control.Applicative
import Control.Arrow
import Control.Monad
import System.IO

main = do
 let f = IOArrow print .
           arr length .
           arr words .
           IOArrow readFile
    -- vs print . length . words . readFile
    runIOArrow f "jabberwocky.txt"
```

Arrow operators

By defining `arr`, `first`, and `(.)` on the `Category` instance, we get many other operators for free:

```
(<<<) :: Category cat => cat b c -> cat a b -> cat a c
```

The `(<<<)` operator defines `Arrow` composition. It the same as the `(.)` function composition defined in the arrow's `Category` instance:

```
(<<<) :: Category cat => cat b c -> cat a b -> cat a c
```

Instead of using `(.)`, we could have composed our arrows with the `(<<<)` operator:

```
main = do
 let f = (IOArrow print) <<< (arr length) <<<
           (arr words) <<< (IOArrow readFile)
    runIOArrow f "jabberwocky.txt"
```

> The `(>>>)` operator is simply composition, reversed:
> ```
> (>>>) = flip (.)
> ```

For example, we could have said this:

```
main = do
let f = (IOArrow readFile) >>>
         (arr words) >>>
         (arr length) >>>
         (IOArrow print)
    runIOArrow f "jabberwocky.txt"
```

To understand the `first` operator, let's tweak our arrow pipeline:

```
main = do
 let f = (IOArrow readFile) >>>
          (arr words) >>>
          (arr (\x -> (x,x))) >>> -- split stream in 2
          (arr length) >>>
          (IOArrow print)
     runIOArrow f "jabberwocky.txt"
     - INVALID code
```

After computing `(arr words)`, we split the result into a 2-tuple, with both parts containing the words in the file. This won't work anymore because `(arr length)` expects to be fed a single value and will now be given a tuple instead.

But we can tell the `(arr length)` arrow to only act on the first element in the tuple:

```
main = do
  let f = (IOArrow readFile) >>> (arr words) >>>
       (arr (\x -> (x,x))) >>>  -- split stream in 2
       (first (arr length)) >>> -- act on first tuple value
       (IOArrow print)
    runIOArrow f "jabberwocky.txt"
```

Now our final result is a tuple, the first part containing the word count, and the second the original words, which remained untouched. In this way, we can create "side channels" in our pipelines, which enable us to share state across arrows in a pipeline. Similarly, the `second` operator works on the second part of the arrow's input:

```
main = do
  let f = (IOArrow readFile)   >>> (arr words) >>>
       (arr (\x -> (x,x)))   >>> -- split stream in 2
       (first (arr length))  >>> -- act on fst tuple value
       (second (arr head))   >>> -- act on snd tuple value
       (IOArrow print)
    runIOArrow f "jabberwocky.txt"
```

Now we're doing two different arrow computations on the different branches. The final operator we'll look at is (***):

```
main = do
  let f = (IOArrow readFile) >>> (arr words) >>>
      (arr (\x -> (x,x))) >>>
      (arr length *** arr head) >>>
      (IOArrow print)
  runIOArrow f "jabberwocky.txt"
```

This operator does what we did earlier with the first and second: it runs two Arrow on the first and second tuple values.

The ability of arrows to take multiple inputs enables us to build intricate pipelines with branching and merging of stream values.

Kleisli arrows and monad arrows

Our IOArrow type:

```
data IOArrow a b = IOArrow {runIOArrow :: a -> IO b}
```

is unnecessary because there is already the Kleisli arrow, which generalizes IOArrow to all Monad:

```
-- (already defined in Control.Arrow)
data Kleisli m a b = K {runKleisli :: a -> m b}
instance Monad m => Arrow (Kleisli m) where
  arr f = K (\x -> return (f x))
  K f >>> K g = K (\x -> f x >>= g)
```

Thanks to Kleisli arrows, we could have just said this:

```
main = do
  let f = Kleisli print . arr length .
      arr words . Kleisli readFile
  runKleisli f "jabberwocky.txt"
```

"This shows that arrows do indeed generalize monads; for every monad type, there is a corresponding arrow type. (Of course, it does not follow that every monadic program can be rewritten in terms of arr and >>>.)"

--John Hughes, Generalising Monads to Arrows

All monads have a corresponding Kleisli arrow, but there are more arrows than monads. In the same way, there are more applicative functors than arrows, and more functors than applicative functors.

> *NOTE: "idiom" means "applicative functor"*
>
> *"idioms embed into arrows and arrows embed into monads"*
>
> *"We have characterized idioms, monads and arrows as variations on a single calculus, establishing the relative order of strength as idiom, arrow, monad..."*
>
> *"Idioms are oblivious, Arrows are meticulous, Monads are promiscuous"*
>
> *–Sam Lindley, Philip Wadler, and Jeremy Yallop*

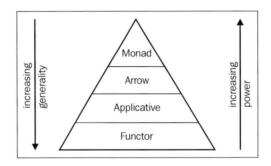

Despite arrows being more generic than monads, we left this pattern until last because it is easier to understand from the perspective of a monad. Also, monads have been around in Haskell (since the early 1990s) for much longer than arrows (since the early 2000s).

We can now add arrows to the list of ways of doing "effectful programming" (IO). Moreover, in the same way that we could compose different monads together into monad stacks, we can combine different arrows into stacks using transformer arrows.

Returning to the abstraction hierarchy, as the types get more powerful (from functor to monad) the means of composition becomes poorer! This is another example of the tradeoff between power and flexibility.

Why arrows?

In 1996, Swierstra and Duponcheel published an optimization to deal with the space leaks prevalent in monadic parsers. Their solution avoided the use of monads.

(S.D. Swierstra, L. Duponcheel. Deterministic, error-correcting combinator parsers, Advanced Functional Programming, Lecture Notes in Computer Science Tutorial, Vol. 1129, Springer, Berlin, 1996, pages 184–207).

John Hughes generalized their ideas, and in the process, generalized monads to arrows. Hughes published the first arrow paper in 2000. It took a few more years for arrows to become established in Haskell.

> "...When libraries emerge which cannot, for fundamental reasons, use the monad interface... Swierstra and Duponcheel have developed a very interesting library for parsing LL-1 grammars ... Yet Swierstra and Duponcheel's optimization is incompatible with the monad interface. We believe that their library is not just an isolated example, but demonstrates a generally useful paradigm for combinator design that falls outside the world of monads."

> "While arrows are a little less convenient to use than monads, they have significantly wider applicability. They can therefore be used to bring the benefits of monad-like programming to a much wider class of applications."

> *–John Hughes, Generalising Monads to Arrows*

Arrows have been useful in parsers, streaming applications, and user interfaces, and in recent years they have featured prominently as an approach to functional reactive programming. (See `https://wiki.haskell.org/Yampa`, `https://wiki.haskell.org/Netwire`).

Summary

This chapter focused on ways of composing functions and types. We saw that functor, applicative, arrow, and monad all lie on a spectrum of generality, with monad being the most powerful type.

We also found how different the historical development of these types were from their natural order in the abstraction hierarchy. This explained the redundancies in type-class methods across these types.

The crux of this chapter was to explore the ways in which function composition can be raised to the level of richer types (than functions).

In the next chapter, we'll explore how "folding" and "traversing" (something we're very used to when working with lists) can be generalized for applicative, arrow, and monad. To get to the heart of folding, we'll take a good look at the very simple and powerful monoid, and see how intricately it is related to folding over data types.

We'll also learn about functional lenses and see how they relate to traversing ad hoc data structures.

4
Patterns of Folding and Traversing

In this chapter, we'll focus on two fundamental patterns of recursion—folding and mapping. The more primitive forms of these patterns are to be found in the Prelude, the "old part" of Haskell.

With the introduction of Applicative came more powerful mapping (traversal), which opened the door to type-level folding and mapping in Haskell. First, we look at how Prelude's list fold is generalized to all Foldable containers. Then, we follow the generalization of the list map to all Traversable containers.

Our exploration of fold and map culminates with the Lens library, which raises Foldable and Traversable to an even higher level of abstraction.

In this chapter, we will cover the following:

- Folding with monoids
- Foldable
- Mapping over lists
- Traversable
- Modernizing Haskell
- Lenses

Folding over lists

Recall the following from *Chapter 1, Functional Patterns – the Building Blocks*:

```
sumLazy [] = 0
sumLazy (x:xs) = x + sumLazy xs
```

This is captured by the recursive process `foldr`:

```
sumLazy' = foldr (+) 0
```

On the other hand, we can write a strict sum function:

```
sumStrict acc [] = acc
sumStrict acc (x:xs) = sumStrict (acc + x) xs
```

This is captured by the iterative process `foldl`:

```
sumStrict' = foldl (+)
```

The `foldl` function is tail recursive and strict in the accumulator. The `foldr` function is non-tail recursive and lazy. In fact, the `foldl` function is a special case of `foldr` (https://wiki.haskell.org/Foldl_as_foldr).

Folding with monadic functions

Let's write a tail-recursive sum that does some IO at each accumulation step:

```
doSumStrict :: (Show a, Num a) => a -> [a] -> IO a

doSumStrict acc [] = return acc
doSumStrict acc (x:xs) = do
  putStrLn $ " + " ++ (show x) ++ " = " ++ (show acc')
  doSumStrict acc' xs
  where acc' = acc + x

main = doSumStrict 0 [2, 3, 5, 7]
```

To write this as a left-fold, we use the `foldM` function:

```
doSumStrict' = foldM doPlus
  where   doPlus acc x = do
    putStrLn $ " + " ++ (show x)
                ++ " = " ++ (show acc)
    return (acc + x)

main = doSumStrict' 0 [2, 3, 5, 7]
```

Both `foldl/r` and `foldM` functions describe folding over lists:

```
foldM :: Monad m => (b -> a -> m b)  -> b -> [a] -> m b
foldl ::             (b -> a -> b)    -> b -> [a] -> b
```

For `foldM`, the fold function is monadic and the result accumulates inside the `Monad` class.

Folding with monoids

Folding allows us to express all manner of accumulations:

```
sum'     = foldr (+)   0
product' = foldr (*)   1
concatS' = foldr (++)  ""
concatL' = foldr (++)  []

any'  = foldr (||) False
all'  = foldr (&&) True

main = do
  print $ sum'      [2, 3, 5]
  print $ product'  [2, 3, 5]
  print $ concatS'  ["2", "3", "5"]
  print $ concatL'  [["2"], ["3"], ["5"]]
  print $ any'      [False, False, True]
  print $ all'      [True, True, True]
```

These accumulation functions differ only in the same function and initial value. In each case, the initial value is also the identity for the corresponding operator:

```
0         +     x = x
1         *     x = x
""        ++    x = x
[]        ++    x = x
False ||        x = x
True  &&    x = x
```

The `Monoid` instance describes exactly this: an associative operator with an identity value:

```
class Monoid a where
    mempty  :: a
    mappend :: a -> a -> a
    mconcat :: [a] -> a
    mconcat = foldr mappend mempty
```

The `mconcat` function is a generic version of our accumulation functions.

Numbers are monoidal under addition and multiplication:

```
newtype Sum' a = Sum' {getSum' :: a} deriving (Show)
instance Num a => Monoid (Sum' a) where
    mempty = Sum' 0
    Sum' x 'mappend' Sum' y = Sum' (x + y)


newtype Product' a = Product' {getProduct' :: a}
instance Num a => Monoid (Product' a) where
    mempty = Product' 1
    Product' x 'mappend' Product' y = Product' (x * y)
```

These are already defined in `Data.Monoid` as `Sum` and `Product`.

This allows us to express + and * as a monoidal accumulation:

```
-- 10 + 7
(Sum 10) `mappend` (Sum 7)
-- Sum {getSum = 17}

-- 10 * 7
(Product 10) `mappend` (Product 7)
-- Product {getProduct = 70}
```

Similarly, `Bool` is a `Monoid` under the `(||)` and `(&&)` operators (corresponding to `Any`, `All`). The `List` instance is a `Monoid` under `(++)` and `[]`, and `String` is a list. This means that we can rewrite our earlier folds in monoidal form:

```
-- import Data.Monoid
main = do
  print $ mconcat [Sum 2, Sum 3, Sum 5]
  print $ mconcat [Product 2, Product 3, Product 5]
  print $ mconcat ["2", "3", "5"]
  print $ mconcat [["2"], ["3"], ["5"]]
  print $ mconcat [Any False, Any False, Any True]
  print $ mconcat [All True, All True, All True]
```

Folding accumulates and monoids are the datatype for accumulation. This is why the generalization of `fold`, described by the Foldable library rests on `Monoid`.

Foldable

We can fold over lists with regular (`foldl/r`) and monadic functions (`foldM`),
but this doesn't help us with folding over any other data structures, for example:

```
data Tree a = Node a (Tree a) (Tree a)
            | Leaf a
   deriving (Show)
```

If we constrain the inner type of `Tree` to be a `Monoid`, we can fold over `Tree`:

```
foldT :: Monoid a => Tree a -> a
foldT (Leaf x) = x
foldT (Node x lTree rTree)
   = (foldT lTree) 'mappend' x 'mappend' (foldT rTree)
     -- using the underlying type's 'mappend'

main = print .foldT $ Node (Sum 2)
                                (Leaf (Sum 3))
                                (Leaf (Sum 5))

print.foldT $ Node (Product 2)
                                (Leaf (Product 3))
                                (Leaf (Product 5))
-- Product {getProduct = 30}
```

It's a pain to inject `Sum` and `Product` into the `Tree` structure. We can simplify things
by passing in a function that will raise the elements being folded over to `Monoid`:

```
foldT' :: Monoid a => (t -> a) -> Tree t -> a
foldT' toMonoid (Leaf x) = toMonoid x
foldT' toMonoid (Node x lTree rTree)
   = (foldT' toMonoid lTree)
     'mappend' (toMonoid x)
     'mappend' (foldT' toMonoid rTree)

main = do
    print $ foldT' Sum            aTree
    print $ foldT' Product        aTree
    print $ foldT' (Any . (==5)) aTree
    print $ foldT' (All . (>0))   aTree
  where aTree = Node 2 (Leaf 3) (Leaf 5)
```

The `Foldable` type-class describes a more general interface for folding:

```
class Foldable (t :: * -> *) where
  -- implement foldMap or foldr
  foldMap :: Monoid m => (a -> m)              -> t a -> m
  foldr ::                       (a -> b -> b)  -> b -> t a -> b
  -- get these for free:
  -- fold, foldl, fold, foldr', foldl', foldr1, foldl1
```

The `foldMap` function generalizes our `foldT'` instance:

```
foldT'  :: Monoid m => (a -> m) -> Tree a -> m
foldMap :: Monoid m => (a -> m) -> t a     -> m
fold    :: Monoid m =>              t m     -> m
```

Also, `fold` assumes that the elements are already monoid:

```
fold = foldmap id
```

Let's make our `Tree` a `Foldable` instance:

```
import Data.Monoid
import qualified Data.Foldable as F
import qualified Control.Monad as M

instance F.Foldable Tree where
  foldMap toMonoid (Leaf x) = toMonoid x
  foldMap toMonoid (Node x lTree rTree)
    = (F.foldMap toMonoid lTree)
      'mappend' (toMonoid x)
      'mappend' (F.foldMap toMonoid rTree)

main = do
 print $ F.foldMap Sum             aTree
 print $ F.foldMap Product         aTree
 print $ F.foldMap (Any . (==5))  aTree
 print $ F.foldMap (All . (>0))    aTree
  where aTree = Node 2 (Leaf 3) (Leaf 5)
```

Instead of just implementing `fold` for `Tree`, we turned `Tree` into a `Foldable` container. The `Data.Foldable` instance comes with many convenience functions that generalize the corresponding Prelude functions:

```
main = do
  print $ F.sum aTree
  print $ F.product aTree
  print $ F.any (==5) aTree
```

```
    print $ F.all (>0) aTree
    print $ F.maximum aTree
    where aTree = Node 2 (Leaf 3) (Leaf 5)
```

Consider the following example, where we can see how List is generalized to the Foldable container for sum:

```
sum    ::              Num a  => [a] -> a
F.sum :: (Foldable t, Num a) => t a -> a
```

In the same way, Foldable.foldM generalizes folding over the Monad class:

```
foldM  :: Monad m  =>
    (b -> a -> m b) -> b -> [a] -> m b

F.foldrM :: (Foldable t, Monad m) =>
    (a -> b -> m b) -> b -> t a -> m b
```

as in

```
doSum = F.foldrM doPlus
  where
    doPlus acc x = do
      putStrLn $ (show x) ++ " = " ++ (show acc)
      return (acc + x)

main = doSum 0 aTree
```

Foldable things can be expressed as lists by folding (:) over the Foldable, for example:

```
main = do
  print $ F.toList aTree
  -- same as
  print $ F.foldr (:) [] aTree
```

This example shows that not all folding accumulates destructively. Fold accumulates the input structure into a single Monoid value, but that single value might be a composite structure. In general, fold transforms a data structure (often radically).

However, both map and filter are special cases of fold, which makes fold the fundamental function of structured recursion.

Foldable generalizes folding over arbitrary data structure and raises the concept of folding to type level. Foldable generalizes structured recursion in Haskell from lists to container types.

Mapping over lists

The `map` function is a specialization of `fold` since we can write map in terms of fold:

```
map f = foldr ((:).f) []
```

Just as with `fold`, we can map over lists with regular or monadic functions:

```
doF n = do print n; return (n * 2)
main = do
    print $ map (* 2) [2, 3, 5, 7]
    mapM  doF [2, 3, 5, 7] >>= print
    mapM_ doF [2, 3, 5, 7]
```

Here:

```
map   :: (a -> b)   -> [a] -> [b]
mapM  :: (a -> m b) -> [a] -> m [b]
mapM_ :: (a -> m b) -> [a] -> m ()
```

In *Chapter 3, Patterns of Composition*, we wrote `mapM` in terms of `sequence` and `sequenceA` (the `Monad` and `Applicative` forms respectively). When we use `sequenceA`, we get a function that maps over `Applicative`:

```
mapA :: Applicative f => (a -> f t) -> [a] -> f [t]
mapA f = sequenceA' . (map f)

sequenceA' :: Applicative f => [f t] -> f [t]
sequenceA' [] = pure []
sequenceA' (x:xs) = (:) <$> x <*> (sequenceA' xs)

-- import Control.Applicative
main = mapA doF [2, 3, 5, 7] >>= print
```

This evaluates as:

```
sequenceA' . (map doF) [2, 3, 5, 7]
(:) <$> (doF 2) <*> sequenceA' . (map doF) [3, 5, 7]
(4:) (:) <$> (doF 3) <*> sequenceA' . (map doF) [5, 7]
...
(4:6:10:14:[])
```

Even though evaluation is lazy, each list element is being visited twice:

- The `mapA` instance traverses the list and applies `f` to each traversed list element
- `sequenceA` performs the resulting actions and re-assembles the results as a list

However, if we define `mapA` in the following way, we will have a single traversal:

```
mapA' f [] = pure []
mapA' f (x:xs) = (:) <$> f x <*> (mapA' f xs)

main = mapA' doF [2, 3, 5, 7] >>= print
```

Given `mapA`, we can define `sequenceA` in terms of it:

```
sequenceA = mapA id
```

This means that `mapA` and `sequenceA` can be defined interdependently:

```
mapA f = sequenceA . (map f)
sequenceA = mapA id
```

Applicative the `map` and `sequence` method are at the heart of the `Traversable` type-class.

Traversable

As with the `Prelude.foldM`, `mapM` fails us beyond lists, for example, we cannot `mapM` over our `Tree` from earlier:

```
main = mapM doF aTree >>= print
-- INVALID
```

The `Traversable` type-class relates to `map` in the same manner as `Foldable` relates to `fold`:

```
-- required: traverse or sequenceA
class (Functor t, Foldable t) => Traversable (t :: * -> *) where
  -- APPLICATIVE form
  traverse  :: Applicative f => (a -> f b) -> t a -> f (t b)
  sequenceA :: Applicative f => t (f a) -> f (t a)

  -- MONADIC form (redundant)
  mapM      :: Monad m    => (a -> m b) -> t a -> m (t b)
  sequence  :: Monad m    => t (m a) -> m (t a)
```

The `traverse` function generalizes our `mapA` function, which was written for lists, to all `Traversable` containers. Similarly, `Traversable.mapM` is a more general version of `Prelude.mapM` for lists:

```
mapM :: Monad m => (a -> m b) -> [a] -> m [b]
mapM :: Monad m => (a -> m b) -> t a -> m (t b)
```

The `Traversable` type-class was introduced along with `Applicative`:

> *"We introduce the type class Traversable, capturing functorial data structures through which we can thread an Applicative computation"*

> --McBride and Paterson, *Applicative Programming with Effects*

A Traversable Tree

Let's make our `Tree` an instance of `Traversable`; we'll start with the difficult way:

```
- a Traversable must also be a Functor and Foldable:
instance Functor Tree where
  fmap f (Leaf x) = Leaf (f x)
  fmap f (Node x lTree rTree)
  = Node (f x)
                (fmap f lTree)
                (fmap f rTree)


instance Foldable Tree where
  foldMap f (Leaf x) = f x
  foldMap f (Node x lTree rTree)
    = (foldMap f lTree)
      'mappend' (f x)
      'mappend' (foldMap f rTree)


--traverse  :: Applicative ma => (a -> ma b) -> mt a -> ma (mt b)
instance Traversable Tree where
  traverse g (Leaf x) = Leaf <$> (g x)
  traverse g (Node x ltree rtree)
    = Node <$> (g x)
      <*> (traverse g ltree) <*> (traverse g rtree)

data Tree a = Node a (Tree a) (Tree a) | Leaf a
  deriving (Show)

aTree = Node 2 (Leaf 3)
                        (Node 5 (Leaf 7)
                                (Leaf 11))

-- import Data.Traversable
main = traverse doF aTree
  where doF n = do print n; return (n * 2)
```

An easier way to do this is to auto-implement the functor, Foldable, and Traversable! (We will see how this magic is done in *Chapter 6, Patterns of Generic Programming*):

```
{-# LANGUAGE DeriveFunctor #-}
{-# LANGUAGE DeriveFoldable #-}
{-# LANGUAGE DeriveTraversable #-}
import Data.Traversable

data Tree a = Node a (Tree a) (Tree a)| Leaf a
  deriving (Show, Functor, Foldable, Traversable)

aTree = Node 2 (Leaf 3)
                       (Node 5 (Leaf 7)
                               (Leaf 11))

main = traverse doF aTree
  where doF n = do print n; return (n * 2)
```

The traversal and the Iterator pattern

The "Gang of Four" Iterator pattern is concerned with providing a way

> *"To access the elements of an aggregate object sequentially without exposing its underlying representation"*

> *--Gamma et al, "Gang of Four" Design Patterns, 1995*

In *The Essence of the Iterator Pattern*, Jeremy Gibbons shows precisely how the Applicative traversal captures the Iterator pattern.

The Traversable.traverse instance is the Applicative version of Traversable. mapM, which means that it is more general than mapM (because Applicative is more general than Monad).

Moreover, because `mapM` does not rely on the monad bind chain to communicate between iteration steps, `Monad` is a superfluous type for mapping with effects (`Applicative` is sufficient). In other words, `traverse` in Applicative is superior to the monadic traversal (`mapM`).

> *"In addition to being parametrically polymorphic in the collection elements, the generic traverse operation is parameterized along two further dimensions: the datatype being traversed, and the Applicative functor in which the traversal is interpreted"*

> *"the improved compositionality of Applicative functors over monads provides better glue for fusion of traversals, and hence better support for modular programming of iterations"*

> *--Jeremy Gibbons, The Essence of the Iterator Pattern*

Modernizing Haskell 98

The introduction of `Applicative`, along with `Foldable` and `Traversable`, have had a big impact on Haskell.

The `Foldable` instance and `Traversable` lift Prelude `fold` and `map` to a much higher level of abstraction. Moreover, `Foldable` and `Traversable` also bring a clean separation between processes that preserve or discard the shape of the structure being processed:

- `Traversable`: This describes a process that preserves that shape of the data structure being traversed over
- `Foldable`: This process, in turn, discards or transforms the shape of the structure being folded over

Since `Traversable` is a specialization of `Foldable`, we can say that shape preservation is a special case of shape transformation. This line between shape preservation and transformation is clearly visible from the fact that functions that discard their result (for example, `mapM_`, `forM_`, `sequence_`, and so on) are in `Foldable`, while their shape-preserving counterparts are in `Traversable`.

Due to the relatively late introduction of `Applicative`, the benefits of `Applicative`, `Foldable`, and `Traversable` only recently found their way into the core of the language.

This change was managed under the "Foldable Traversable in Prelude" proposal (planned for inclusion in the core libraries from GHC 7.10), which can be found at https://wiki.haskell.org/Foldable_Traversable_In_Prelude.

The proposal involves replacing less generic functions in `Prelude`, `Control.Monad`, and `Data.List` with their more polymorphic counterparts in `Foldable` and `Traversable`.

There have been objections to this modernization, the main concern being that more generic types are harder to understand and that this may compromise Haskell as a learning language. While these valid concerns will indeed have to be addressed, it has not prevented the Haskell community from climbing to new abstract heights.

Lenses

A Lens provides access to a particular part of a data structure.

Lenses express a high-level pattern for composition and in that sense belong firmly in *Chapter 3, Patterns for Composition*. However, Lens is also deeply entwined with Foldable and Traversable, and so we describe it in this chapter instead.

Lenses relate to the getter and setter functions, which also describe access to parts of data structures. To find our way to the Lens abstraction (as per Edward Kmett's Lens library), we'll start by writing a getter and setter to access the root node of a tree.

Deriving Lens

Let's return to our tree from earlier:

```
data Tree a = Node a (Tree a) (Tree a)
                | Leaf a
  deriving (Show)

intTree
= Node 2 (Leaf 3)
             (Node 5 (Leaf 7)
                         (Leaf 11))

listTree
 = Node [1,1] (Leaf [2,1])
                     (Node [3,2] (Leaf [5,2])
                                      (Leaf [7,4]))

tupleTree
 = Node (1,1)
             (Leaf (2,1))
             (Node (3,2)
                        (Leaf (5,2))
                        (Leaf (7,4)))
```

Let's start by writing generic getter and setter functions:

```
getRoot :: Tree a      -> a
getRoot (Leaf z)       = z
getRoot (Node z _ _) = z

setRoot :: Tree a -> a -> Tree a
setRoot (Leaf z)       x = Leaf x
setRoot (Node z l r) x = Node x l r

main = do
  print $ getRoot intTree
  print $ setRoot intTree 11
  print $ getRoot (setRoot intTree 11)
```

If, instead of setting a value, we want to pass in a setter function, use the following code:

```
fmapRoot :: (a -> a) -> Tree a -> Tree a
fmapRoot f tree = setRoot tree newRoot
  where newRoot = f (getRoot tree)
```

We have to get the root of the Tree, apply the function, and then set the result. This double work is akin to the double traversal we saw when writing traverse in terms of sequenceA. In that case, we resolved the issue by defining traverse (followed by sequenceA in terms of traverse).

We can do the same thing here, by writing fmapRoot to work in a single step (and then rewriting setRoot' in terms of fmapRoot'):

```
fmapRoot' :: (a -> a) -> Tree a -> Tree a
fmapRoot' f (Leaf z)       = Leaf (f z)
fmapRoot' f (Node z l r) = Node (f z) l r

setRoot' :: Tree a -> a -> Tree a
setRoot' tree x = fmapRoot' (\_ -> x) tree

main = do
  print $ setRoot' intTree 11
  print $ fmapRoot' (*2) intTree
```

The `fmapRoot'` function delivers a function to a particular part of the structure and returns the same structure:

```
fmapRoot' :: (a -> a) -> Tree a -> Tree a
```

To me a provision for IO, we need a new function:

```
fmapRootIO :: (a -> IO a) -> Tree a -> IO (Tree a)
```

We can generalize this beyond IO to all monads:

```
fmapM :: (a -> m a) -> Tree a -> m (Tree a)
```

It turns out that if we relax the requirement for `Monad` and instead generalize to all the `Functor` container types `f'`, then we get a simple van Laarhoven Lens of type:

```
type Lens' s a
    = Functor f' =>
            (a -> f' a) -> s -> f' s
```

The remarkable thing about a van Laarhoven Lens is that, given the above function type, we also gain `get`, `set`, `fmap`, and `mapM`, along with many other functions and operators.

The `Lens` function type signature is all that it takes to make something a Lens that can be used with the `Lens` library. It is unusual to use a type signature as the "primary interface" for a library. The immediate benefit is that we can define a Lens without referring to the `Lens` library.

We'll explore more benefits and costs to this approach, but first let's write a few lenses for our tree.

(The derivation of the Lens abstraction used here is based on Jakub Arnold's "Lens tutorial", which is available at http://blog.jakubarnold.cz/2014/07/14/lens-tutorial-introduction-part-1.html.)

Writing a Lens

A Lens is said to provide focus on an element in a data structure.

Our first Lens will focus on the root node of a `Tree` instance. Using the Lens type signature as our guide, we arrive at the following:

```
lens':: Functor f  => (a -> f' a) -> s       -> f' s
root :: Functor f' => (a -> f' a) -> Tree a -> f' (Tree a)
```

This is still not very tangible; fmapRootIO is easier to understand, with the Functor f' being IO:

```
fmapRootIO :: (a -> IO a) -> Tree a -> IO (Tree a)
fmapRootIO g (Leaf z)    = (g z) >>= return . Leaf
fmapRootIO g (Node z l r) = (g z) >>= return . (\x -> Node x l r)

displayM x = print x >> return x

main = fmapRootIO displayM intTree
```

If we drop down from Monad into Functor, we have a Lens for the root of a Tree:

```
root :: Functor f' => (a -> f' a) -> Tree a -> f' (Tree a)
root g (Node z l r) = fmap (\x -> Node x l r) (g z)
root g (Leaf z)     = fmap Leaf            (g z)
```

As monad is a functor, this function also works with monadic functions:

```
main = root displayM intTree
```

As root is a Lens, the Lens library gives us the following:

```
-- import Control.Lens
main = do
  -- GET
  print $ view root listTree
  print $ view root intTree
  -- SET
  print $ set root [42] listTree
  print $ set root 42   intTree
  -- FMAP
  print $ over root (+11) intTree
```

The over instance is the way in which Lens uses fmap to map a function into a Functor method.

Composable getters and setters

We can have another Lens on a Tree instance to focus on the rightmost leaf:

```
rightMost :: Functor f' =>
  (a -> f' a) -> Tree a -> f' (Tree a)

rightMost g (Node z l r)
  = fmap (\r' -> Node z l r') (rightMost g r)
rightMost g (Leaf z)
  = fmap (\x -> Leaf x) (g z)
```

The `Lens` library provides several lenses for `Tuple` (for example, `_1` which brings focus to the first `Tuple` element). We can compose our `rightMost` Lens with the `Tuple` lenses:

```
main = do
  print $ view rightMost tupleTree
  print $ set rightMost (0,0)  tupleTree

  -- Compose Getters and Setters
  print $ view (rightMost._1) tupleTree
  print $ set (rightMost._1) 0 tupleTree
  print $ over (rightMost._1) (*100) tupleTree
```

A Lens can serve as a getter, setter, or modifier.

Moreover, we compose Lenses using the regular function composition `(.)`! Note that the order of composition is reversed: in `(rightMost._1)`, the `rightMost` Lens is applied before the `_1` Lens.

Lens Traversal

A Lens focuses on one part of a data structure, not several; for example, a Lens cannot focus on all the leaves of a `Tree`:

```
set leaves 0 intTree
over leaves (+1) intTree
```

To focus on more than one part of a structure, we need a `Traversal`, the Lens generalization of `Traversable`. As Lens relies on `Functor`, `Traversal` relies on `Applicative`. Other than this, the signatures are exactly the same:

```
traversal :: Applicative f' =>
    (a -> f' a) -> Tree a -> f' (Tree a)
lens :: Functor f'=>
    (a -> f' a) -> Tree a -> f' (Tree a)
```

A `leaves` `Traversal` delivers the setter function to all the leaves of the `Tree`:

```
leaves :: Applicative f' => (a -> f' a) -> Tree a -> f' (Tree a)
leaves g (Node z l r)
  = Node z <$> leaves g l <*> leaves g r
leaves g (Leaf z)
  = Leaf <$> (g z)
```

We can use set and over with our new `Traversal` instance:

```
set leaves 0 intTree
over leaves (+1) intTree
```

Traversals compose seamlessly with `Lenses`:

```
main = do
  -- Compose Traversal + Lens
  print $ over (leaves._1) (*100) tupleTree

  -- Compose Traversal + Traversal
  print $ over (leaves.both) (*100) tupleTree

  -- map over each elem in target container (e.g. list)
  print $ over (leaves.mapped) (*(-1)) listTree

  -- Traversal with effects
  mapMOf leaves displayM tupleTree
```

(The `both` function is a Tuple Traversal that focuses on both elements).

Lens.Fold

The `Lens.Traversal` function lifts `Traversable` into the realm of lenses, while `Lens.Fold` does the same for `Foldable`:

```
main = do
  print $ sumOf leaves intTree
  print $ anyOf leaves (>0) intTree
```

Just as for `Foldable` sum and `foldMap`, we can write Lens `sumOf` in terms of `foldMapOf`:

```
getSum $ foldMapOf lens Sum
```

where `foldMapOf` is a generalization of `Foldable.foldMap`.

The Lens library

We've used only simple Lenses; a fully parameterized Lens would allow for replacing parts of a data structure with different types:

```
type Lens s t a b = Functor f' => (a -> f' b) -> s -> f' t
-- vs simple Lens
type Lens' s a = Lens s s a a
```

Lens library function names do their best not to clash with existing names; for example, postfixing of idiomatic function names with "Of" (sumOf, mapMOf, and so on), or using different verb forms such as droppingWhile instead of dropWhile. While this creates a burden in terms of having to learn new variations, it does have a big plus: allowing for easy unqualified import of the Lens library.

By leaving the Lens function type transparent (and not obfuscating it with a newtype), we get Traversals by simply swapping out functor for Applicative in the Lens type definition. We also get to define lenses without having to reference the Lens library. On the downside, Lens type signatures can be bewildering at first sight. They form a language of their own that requires an effort to get used to, for example:

```
mapMOf :: Profunctor p =>
  Over p (WrappedMonad m) s t a b -> p a (m b) -> s -> m t

foldMapOf :: Profunctor p =>
  Accessing p r s a -> p a r -> s -> r
```

On the surface, the Lens library gives us composable getters and setters, but there is much more to lenses than that. By generalizing Foldable and Traversable into Lens abstractions, the Lens library lifts getters, setters, lenses, and Traversals into a unified framework in which they all compose together.

The Lens library has been criticized for not reflecting idiomatic Haskell and for simply taking on too much responsibility. Nevertheless, Edward Kmett's Lens library is a sprawling masterpiece that is sure to leave a lasting impact on Haskell.

Summary

We followed the evolution of fold and map, starting with lists (Haskel 98), then generalizing for all Foldable/Traversable containers (in the mid 2000s).

Following that, we saw how the Lens library (2012) places folding and traversing in an even broader context. Lenses give us a unified vocabulary to navigate through data structures, which explains why it has been described as a "query language for data structures".

In this chapter, as we moved up the layers of abstraction, the function type signatures became ever more generic.

Genericity through type parameterization represents only one kind of generic programming. In *Chapter 6, Patterns of Generic Programming,* we explore more such patterns. But before we can do that, we need to add a few more advanced language extensions to our toolset. This is the subject of the next chapter.

5

Patterns of Type Abstraction

In the previous chapter, we explored several ways in which the Haskell 98 language has been evolving via the new libraries. This chapter brings to focus some key advances on another major front of Haskell's evolution: language extensions.

> *"Haskell's type system has developed extremely anarchically. Many of the new features were sketched, implemented, and applied well before they were formalized. This anarchy has both strengths and weaknesses. The strength is that the design space is explored much more quickly the weakness is that the end result is extremely complex, and programs are sometimes reduced to experiments to see what will and will not be acceptable to the compiler."*
>
> *--Hudak et al., History of Haskell*

Haskell extensions are tied to the compiler implementation rather than the language standard (the GHC compiler is our reference compiler). We'll explore the extensions along three axes: abstracting functions, datatypes, and type-classes.

- **Abstracting function types**: RankNTypes
- **Abstracting datatypes**: Existential quantification, phantom types, **generalized algebraic datatypes (GADTs)**, type case pattern, dynamic types, and heterogeneous lists
- **Abstracting type-classes**: Multiparameter type-classes and functional dependencies

Abstracting function types: RankNTypes

Consider the higher order function that maps the argument function to each
tuple element:

```
tupleF elemF (x, y) = (elemF x, elemF y)
```

Left to its own devices, the Haskell 98 compiler will infer this type for a `tupleF`
function:

```
tupleF :: (a -> b) -> (a, a) -> (b, b)
```

As `elemF` is applied to x and y, the compiler assumes that x and y must be of the
same type, hence the inferred tuple type `(a, a)`. This allows us to do the following:

```
tupleF length ([1,2,3], [3,2,1])
tupleF show (1, 2)
tupleF show (True, False)
```

However, not this:

```
tupleF show (True, 2)
tupleF length ([True, False, False], [1, 2, 4])
```

RankNTypes allow us to enforce parametric polymorphism explicitly. We want
`tupleF` to accept a polymorphic function of arguments; in other words, we want
our function to have a "higher rank type", in this case `Rank 2`:

```
{-# LANGUAGE Rank2Types #-}

tupleF' :: (Show a1, Show a2) =>
  (forall a . Show a => a -> b) -- elemF
    -> (a1, a2) -> (b, b)

tupleF' elemF (x, y) = (elemF x, elemF y)
```

The use of `forall` in the `elemF` function signature tells the compiler to make `elemF`
polymorphic in a, as shown:

```
main = do
  -- same as before...
  print $ tupleF' show (1, 2)
  print $ tupleF' show (True, False)

  -- and now we can do this...
  print $ tupleF' show (True, 2)
```

We can see the polymorphism clearly in the last line of the preceding code, where `show` is applied polymorphically to the types `Bool` and `Int`.

The rank of a type refers to the nesting depth at which polymorphism occurs. `Rank 0` describes absence of polymorphism. For example:

```
intAdd :: Int -> Int -> Int
```

`Rank 1` refers to regular parametric polymorphism, for example:

```
id :: a -> a
```

while `tupleF'`, `Rank 2` to `Rank n` describes a deeper nested polymorphism. However, the deeper the level of polymorphism, the rarer and more exotic the application. Having said that,

> *"These cases are not all that common, but there are usually no workarounds; if you need higher-rank types, you really need them!"*
>
> -- *Peyton Jones et al., 2007, Practical type inference for arbitrary-rank types*

Abstracting datatypes

In this section, we will describe a series of patterns related to data abstraction. We start with existentially quantified types then progress to phantom types and end with GADTs. We'll see that these patterns are based on a spectrum of generality and power.

Universal quantification

Let's explore existential quantification from the perspective of its opposite, universal quantification. We rely on universal quantification whenever we parameterize function types, for example:

```
id' ::            a -> a
-- is the same as
id' :: forall a. a -> a

id' x = x
```

In general, universal quantification expresses parametric polymorphism in functions and datatypes. We use the `forall` keyword in `Rank-n` function type to indicate nested parametric polymorphism. Similarly, universal quantification is the default pattern when parameterizing types with types, as shown:

```
data Maybe' a =            Nothing' | Just' a
-- conceptually (but not practically!) the same as
data Maybe' a = forall a. Nothing' | Just' a
```

As another example, consider the following universally qualified type:

```
data ObjU a
  = ObjU a                  -- property
            (a -> Bool)   -- obj method
            (a -> String) -- obj method
```

Here, we mimic an object with the property of type a and two object methods. We can apply a method to the property by extracting the value and the method with pattern matching:

```
obj_f1 :: ObjU a -> Bool
obj_f1 (ObjU v f1 _) = f1 v

obj_f2 :: ObjU a -> String
obj_f2 (ObjU v _ f2) = f2 v

main = do
  print $ obj_f1 obj -- even 3
  print $ obj_f2 obj -- show 3
  where obj = (ObjU 3 even show)
```

We've packaged a value with some functions in the `ObjU` object but we haven encapsulated the value as a true object would have. This is what existential quantification enables us to do.

Existential quantification and abstract datatypes

The existentially qualified `ObjE` object hides the type parameter instead of the universally qualified `ObjU` object, which is no longer present on the left-hand side (although it's confusing, we use the `forall` keyword in both cases):

```
--   ObjU a = forall a. ObjU a (a -> Bool) (a -> String)
data ObjE  = forall a. ObjE a (a -> Bool) (a -> String)
```

This means that the type parameter is also "hidden" in the type signature of `objE_f1,2`, as shown in the following code:

```
objE_f1 :: ObjE -> Bool
objE_f1 (ObjE v f1 _) = f1 v

objE_f2 :: ObjE -> String
objE_f2 (ObjE v _ f2) = f2 v

-- requires {-# LANGUAGE ExistentialQuantification #-}
main = do
  print $ objE_f1 obj -- even 3
  print $ objE_f2 obj -- show 3
  where obj = (ObjE 3 even show)
```

We can access an encapsulated object property only with the functions packaged with that property. For example, we can't apply any other function to the following property:

```
-- INVALID (cannot infer types)
objE_f3 (ObjE v f1 f2) = v
```

Existential quantification provides us with the means to implement abstract datatypes, thus providing functions over a type while hiding the representation of the type. We can also achieve the abstract datatypes on a higher level by using Haskell's module system to hide algebraic datatype constructors while exporting functions over the type.

For the remainder of this chapter, we will refer to types that are existentially qualified as "existentials".

Universal	Existential
Type parametrization	Type abstraction
Parametric Polymorphism	Encapsulation
user of data specifies type	implementer of data specifies type
forall = "for all"	forall = "for some"

Phantom types

Phantom types were introduced in 1999 as a solution to the challenges that arise when embedding a type-safe **domain specific language** (DSL) in Haskell. (See *Fun with Phantom Types*, Hinze http://www.cs.ox.ac.uk/ralf.hinze/publications/With.pdf)

Consider this trivial expression language and evaluator:

```
data Expr1 = I1 Int
                | Add1 Expr1 Expr1

eval1 :: Expr1 -> Int
eval1 (I1 v) = v
eval1 (Add1 x y) = (eval1 x) + (eval1 y)
```

When we add another base type (B2 Bool) to the expression language, the situation gets more complicated.

```
data Expr2 = I2 Int
                | B2 Bool
                | Add2 Expr2 Expr2
        deriving Show
```

This has brought about the following two problems:

1. Add2 was only meant to work with I2 Ints, but we can now construct a bad value. The regular algebraic datatypes don't allow us to express this relationship (constraint) between the two constructors:

   ```
   -- construct a "bad" value
       (Add2 (I2 11) (B2 True))
   ```

2. The type inference can no longer be inferred (or defined) for eval, as shown:

   ```
       -- INVALID
       eval2 :: Expr2 -> t
       eval2 (I2 v) = v
       eval2 (B2 v) = v
       eval2 (Add2 x y) = (eval2 x) + (eval2 y)
   ```

In our case the phantom types solve the first problem by adding the type t in

```
data Expr3 t = I3 Int
                | B3 Bool
                | Add3 (Expr3 Int) (Expr3 Int)
    deriving Show
```

The type t serves as a type placeholder that can be used by each constructor to describe its particular type. However, all the constructors still return the same type:

```
I3    ::                              Int -> Expr3 t
B3    ::                             Bool -> Expr3 t
Add3 :: Expr3 Int -> Expr3 Int -> Expr3 t
```

The `Expr3` value is parametrized by the type `t`, but `t` does not appear in any of the constructors, hence the term phantom type. We can still construct invalid values, as shown:

```
Add3 (I3 11) (B3 True)
```

However, we can still use the phantom type information to create type-safe smart constructors:

```
i3 :: Int -> Expr3 Int
i3 = I3

b3 :: Bool -> Expr3 Bool
b3 = B3

add3 :: Expr3 Int -> Expr3 Int
              -> Expr3 Int
add3 = Add3
```

If we use the smart constructors instead of the datatype constructors, the Haskell type-checker will prevent us from creating invalid values. For example, the following will be rejected:

```
-- INVALID
add3 (i3 10) (b3 True)
```

However, type inference remains a problem because the values are still not described accurately. For example:

```
(I3 12) :: Expr3 t    -- this
  --         :: Expr3 Int - not this
```

The effect is that adding values remains too ambiguous:

```
eval3 (Add3 x y) = (eval3 x) + (eval3 y)
```

Despite these limitations, phantom types continue to be useful in other areas. For example, the `Lens` library uses the const phantom type to great effect (See https://github.com/ekmett/lens/wiki/Derivation).

We have seen here how Phantom types enable type-syfe construction (stated as 'problem 1' earlier in this section). To solve problem (2) we'll need the power of generalized algebraic datatypes (GADTs).

Generalized algebraic datatypes

GADTs emerged independently from both ML and Haskell camps in 2003 and were a part of the GHC by 2005 (see *History of Haskell*, Hudak et al. and *First-class Phantom Types*, Cheney and Hinze).

GADTs bring together phantom types, smart constructors, and refined pattern matching:

```
{-# LANGUAGE GADTs #-}

data Expr t where  -- phantom 't'
    -- built-in smart constructors
    I :: Int  -> Expr Int
    B :: Bool -> Expr Bool
    Add :: Expr Int -> Expr Int -> Expr Int
```

The GADT smart constructors describe constrained instances of `Expr t`. As with phantoms types, smart constructors provide increased type safety for data construction. However, GADTs give us something we don't get from phantom types:

```
eval :: Expr t -> t
eval (I v) = v
eval (B v) = v
eval (Add x y) = (eval x) + (eval y)

--  eval (Add (I 10) (I 12))
```

This is because GADT smart constructors are built into the type and we can match the pattern on them. This solves the problem of type inference that we had with phantom types. This is why GADTs are also known as **first-class phantom types**.

GADTs are not expressed by syntax but rather by the relationship between the type parameters and the constructor return types. Similarly, phantom types are not expressed by syntax but implied by the lack of appearance of a type parameter in the type constructors.

There is a subtle drift in the meaning of a type parameter, from signifying the type of some embedded value to expressing the type metadata.

Typecase pattern

Generic programming (the focus of the following *Chapter 6, Patterns of Generic Programming*) is another important use-case for GADTs. As an example, consider a type representation Rep that unifies the three variable types Int, Char, and List:

```
data Rep t where
  RInt  :: Rep Int
  RChar :: Rep Char
  RList :: Show a => Rep a -> Rep [a]
```

The RList function can be thought of as being existentially qualified (a does not appear on the left-hand side). The phantom t in Rep t will serve as the type metadata.

We can now write a function that takes a value along with its type representation:

```
showT :: Show t => Rep t -> t -> String

showT RInt i  = (show i) ++ " :: INT"
showT RChar i = (show i) ++ " :: Char"

showT (RList rep) [] = "THE END"
showT (RList rep) (x:xs)
  = (showT rep x) ++ ", " ++
    (showT (RList rep) xs)
```

The showT function is a type-indexed function because it is defined for each member of the family of types Rep t:

```
showT RInt 3
showT (RList RInt) [12,13,14]
showT (RList RChar) ['2','3','5']
```

More precisely, showT is a closed type-indexed function because the type index family (Rep t) is fixed.

In contrast, the show function of the Show type-class is an example of an open type-indexed function. show is simply type-indexed by instances of Show and considered "open" because we can add new types to the type index freely (in this case, new instances of the type-class Show).

In languages that allow us to reflect on the type of a value, we can write `showT` by dealing with the parameter value on a "type-case" basis:

```
- pseudo code
case (type t) of
  Int  -> (show t) ++ ":: INT"
  Char -> (show t) ++ ":: CHAR"
  List -> -- ... show list
```

This style has been distilled into a design pattern:

> *"TypeCase: a design pattern that allows the definition of closed type-indexed functions, in which the index family is fixed but the collection of functions is extensible"*
>
> *-- Oliveira and Gibbons, TypeCase: A Design Pattern for Type-indexed Functions*

Dynamic types

We now have the ingredients to define dynamic types. All we need to do is package a type together with the type representation, as shown:

```
data DynamicEQ = forall t. Show t =>
                              DynEQ (Rep t) t
```

Here, we've done the packaging using existential quantification. Even though `DynEq` dynamic values have opaque type, they are well typed. For example, we can use them to express heterogeneous lists, as shown:

```
dynEQList = [DynEQ RChar 'x',
             DynEQ RInt 3]
```

Since GADTs generalize existentials, we can also write a "dynamic GADT", such as the following:

```
data Dynamic where
  Dyn :: Show t => Rep t -> t -> Dynamic

instance Show Dynamic where
  show (Dyn rep v) = showT rep v
```

We can use this GADT to define a heterogeneous list of dynamically typed values:

```
dynList :: [Dynamic]
dynList = [Dyn RChar 'x', Dyn RInt 3]

showDyn (Dyn rep v) = showT rep v
```

The showDyn function acts on dynamic values while the generic function showT acts on "generic data".

To achieve representable lists of dynamic types, (RList RDyn), we need to add another constructor to our representation Rep:

```
data Rep t where
  RInt  :: Rep Int
  RChar :: Rep Char
  RList :: Show a => Rep a -> Rep [a]
  RDyn  :: Rep Dynamic
```

as well as another clause for showT to deal with dynamic values (analogous to ShowDyn):

```
showT RDyn (Dyn rep v) = showT rep v
```

Now we have the generic functions acting on dynamic types:

```
main = do
  print $ showT RInt 17
  print $ showT (RList RInt) [12,13,14]
  print $ showT (RList RDyn) dynList
```

Dynamic types carry enough type information about themselves to enable safe type casting:

```
toInt :: Dynamic -> Maybe Int
toInt (Dyn RInt i) = Just i
toInt (Dyn _ _)    = Nothing
```

(See *Fun with Phantom Types*, Hinze: http://www.cs.ox.ac.uk/ralf.hinze/publications/With.pdf,

and *Generalized Algebraic Data Types in Haskell*, Anton Dergunov: https://themonadreader.files.wordpress.com/2013/08/issue221.pdf)

Heterogeneous lists

Earlier in this section, we saw that the GADTs generalize phantom types as well as existentials. To see this, let's explore the heterogeneous lists pattern (lists of varying types).

Using existentials

We can quite easily define a heterogeneous list using existentials:

```
{-# LANGUAGE ExistentialQuantification #-}

data LI_Eq1 = forall  a. LI_Eq1 a

hListEq1 :: [LI_Eq1]
hListEq1 =  [LI_Eq1 3, LI_Eq1 "5"]
```

However, as we saw earlier, we can't do anything with this list. For example, in order to show list items we need to package a show function with each item:

```
data LI_Eq2 = forall  a. LI_Eq2 a (a -> String)

hListEq2 :: [LI_Eq2]
hListEq2 =  [LI_Eq2  3  (show :: Int -> String),
                LI_Eq2 "5" (show :: String -> String)]
```

(We add the show type signatures here for the sake of clarity but they can be inferred, in other words omitted.)

```
showEq2 (LI_Eq2 v showF) = showF v
-- e.g. main = mapM_ (putStrLn . showEq2) hListEq2
```

Passing in the show functions explicitly is a chore that we can do without. Using type-classes (in this case, Show) makes the code more compact because instead of embedding functions, we can apply them by constraining them to type-classes:

```
data LI_Eq3 = forall a. Show a => LI_Eq3 a

hListEq3 :: [LI_Eq3]
hListEq3 =  [LI_Eq3  3, LI_Eq3 "5"]

showEq3 (LI_Eq3 v) = show v
```

The type-class constraint specified in the existential amounts to what is called **bounded quantification** (bounded by type-class).

Using GADTs

We can express heterogeneous lists in the same two styles that we used with existentials. In the first style we pass in the show function (`a -> String`), as shown:

```
{-# LANGUAGE GADTs #-}

data LI_Gadt1 where
  {MkShow1 :: a -> (a -> String) -> LI_Gadt1}

hListGadt1 :: [LI_Gadt1]
hListGadt1 = [MkShow1 "3" show, MkShow1 5 show]

showGadt1 (MkShow1 v showF) = showF v
```

Alternatively, we can also use a GADT with bounded quantification:

```
data LI_Gadt2 where
  {MkShow2 :: Show a => a -> LI_Gadt2}

hListGadt1 :: [LI_Gadt1]
hListGadt2 = [MkShow2 "3", MkShow2 5]

showGadt2 (MkShow2 v) = show v
```

Abstracting type-classes

There are several ways in which type-classes can be generalized further. In this section, we will focus on extending the number of type parameters from one to many. The extension to multiparameter type-classes demands that we specify relations between type parameters by way of functional dependencies.

Multiparameter type-classes

We can view regular type-classes (`for example a`, `Ord a`, `Monad a`, and so on.) as a way to specify a set of types. Multiparameter classes, on the other hand, specify a set of type relations. For example, the `Coerce` type-class specifies a relation between two type parameters:

```
class Coerce a b where
  coerce :: a -> b

instance Coerce Int String where
```

```
coerce = show

instance Coerce Int [Int] where
  coerce x = [x]
```

The type signature of `coerce` is as follows:

```
coerce :: Coerce a b => a -> b
```

This states that `coerce` is the function a -> b if a is coerce-able to b, that is, if the relation (Coerce a b) exists. In our case, `coerce` will work for (Int -> String) and (Int -> [Int]).

However, with multiple type parameters, type inference suffers; for example, the compiler rejects.

```
coerce 12 :: String
```

We have to help it along with type annotations:

```
coerce (12::Int) :: String
coerce (12::Int) :: [Int]
```

This sort of type ambiguity quickly gets out of hand and was the reason for multiparameter type-classes not making it into Haskell 98, despite being part of the GHC since 1997.

Functional dependencies

It was only with the discovery of functional dependencies that multiparameter classes became more practically useful. (Mark Jones, 2000, *Type Classes with Functional Dependencies*: http://www.cs.tufts.edu/~nr/cs257/archive/mark-jones/fundeps.ps).

Functional dependencies give us a way to constrain the ambiguity created by multiple type parameters. For example, we can constrain the relationship between a and b in `Coerce` with a functional dependency:

```
{-# LANGUAGE FunctionalDependencies #-}

class Coerce2 a b | b -> a where
  coerce2 :: a -> b

instance Coerce2 Int String where
  coerce2 = show

instance Coerce2 Int [Int] where
  coerce2 x = [x]
```

The relation (b -> a) tells the compiler that if it can infer b, it can simply look up the corresponding a in one of the type-class instances, for example:

```
coerce2 12 :: String
```

The compiler can infer b :: String and can find the (uniquely) corresponding a :: Int in.

```
instance Coerce2 Int String where ...
```

Moreover, the compiler can now prevent us from adding conflicting instance declarations, for example:

```
-- INVALID
instance Coerce2 Float String where
    coerce2 = show
```

This will mean that b :: String could imply either Int or Float. The functional dependency b -> a tells the compiler that b determines a uniquely.

Summary

This chapter explored some key language extensions and the design patterns associated with them. We followed abstraction along three contours: functions, datatypes, and type-classes, and we found that language extensions come with two major costs:

- impaired type inference (requiring more type annotations)
- affinity to compiler implementations (decreasing portability of code across compilers)

Hence, the sentiment:

> *"Whenever you add a new feature to a language, you should throw out an existing one (especially if the language at hand is named after a logician)"*

> -- *Fun with Phantom Types, Hinze*

We are now well prepared to explore the next chapter.

6
Patterns of Generic Programming

In this chapter, we seek a unified view of "Generic Programming", which comes in many guises; hence the statement,

"Genericity is in the eye of the beholder"

-- Jeremy Gibbons, Datatype-Generic Programming

We start with a broad perspective by reviewing Jeremy Gibbons' patterns of generic programming—many of which we have already encountered. Then we shift focus to one of the patterns—datatype-generic programming—which is characterized by generic functions parameterized by the shape of the datatype instead of the content.

To get a taste of datatype-generic programming, we'll sample three basic approaches: the sum of products, origami programming, and scrapping your boilerplate.

Along the way, we'll encounter a few exotic Haskell types (`Typeable` and `Data`; `Bifunctor` and `Fix`), reveal the underpinnings of Derivable type classes and also discover four Gang of Four design patterns in the heart of datatype-generic programming:

- Patterns of generic programming
- Sum of products style
- Origami programming
- Scrap your boilerplate

Patterns of generic programming

Jeremy Gibbons describes seven patterns of generic programming, where each pattern is viewed as a different kind of parameterization. For more information, refer to his article, *Datatype-Generic Programming*, which is available at http://www.cs.ox.ac.uk/jeremy.gibbons/publications/dgp.pdf.

Patterns 1 and 2 – functions

Functions parameterized by values are more general than functions with hardcoded values. This is the most simple kind of generic programming.

Functions parameterized by functions (higher order functions) represent a more powerful form of genericity.

Pattern 3 – polymorphic types and functions

Types parameterized by other types are as follows:

```
Tree a = Leaf a | Node a (Tree a)
- is more generic than
TreeI  = Leaf Int | Node Int TreeI
```

Functions parameterized by polymorphic types are as follows:

```
f :: Tree a → a
- is more generic than
g :: String → String
```

Pattern 4 – type-class polymorphism

In the previous chapter, we saw that the `Foldable` class provides a uniform interface for folding over different types. For a type to participate in the `Foldable` class, we need to implement some functions specific to the type (for example, `foldMap` or `foldr`).

On the one hand, the type class serves as a contract for ad hoc implementations, while on the other hand, it facilitates generic functions. Through this combination of type-specific and generic functions, type classes give us a way to generalize functions over disparate types. This can be referred to as **ad hoc datatype genericity**.

Pattern 5 – meta-programming

This pattern refers to programs specified by other programs, which can be manifested in many different ways; for example, it can refer to reflection-based programming (analyzing the structure of code and data), template-based meta-programming (for example, `TemplateHaskell`, 2002), or other styles of code generation.

The Derivable type-classes

The Derivable type-classes enable code generation for type-class instances. Haskell 98 included autoderivation for the `Eq`, `Ord`, `Enum`, `Bounded`, `Show`, and `Read` type classes.

A second generation of Derivable type-classes was added to the **Glasgow Haskell Compiler (GHC)** in stages for `Data`, `Typeable`, `Functor`, `Foldable`, `Traversable`, and `Generic`.

Haskell 98 Derivable type-classes were achieved through compiler analysis of the structure of the derivable datatypes, that is, a form of metaprogramming. As there was no unifying formalism underlying the different types, the early Derivable type-classes represent only a rudimentary form of generic programming.

Later in this chapter, we will encounter more generic approaches to enable Derivable type-classes.

Generalized newtype deriving

The `GeneralizedNewtypeDeriving` language extension allows a `newtype` declaration to inherit some or all of the type-class instances of an inner type. This is achieved through a trivial kind of meta-programming.

We used this extension when we created a `Monad` transformer stack:

```
{-# LANGUAGE GeneralizedNewtypeDeriving #-}

newtype App a = App {runApp :: ReaderT Config (Writer String) a}
    deriving (Monad, MonadReader Config, MonadWriter String)
```

Pattern 6 – type laws

Many fundamental datatypes, such as `Functor`, `Applicative`, `Arrow`, and `Monad` are associated with mathematical laws that are meant to be obeyed by the type implementer. Since the Haskell type system is not strong enough to express type laws in general, they are not enforceable by the compiler, and the "burden of proof" remains with the type-class implementer.

There are languages (for example, Coq, Agda, and Idris) with type systems designed for expressing laws and constraints for types (known as **dependently-typed programming languages**; refer to *Chapter 7, Patterns of Kind Abstraction*, for more on this).

This limitation in Haskell (and most languages for that matter) is a strong motivator for Derivable type classes. Generic implementations of type-classes still have to obey type laws, but at least we can get this right in the generic code, and implementers can assume that the type laws are obeyed.

Pattern 7 – datatype generic programming

Although datatype generic programming is a sophisticated pattern, it is based on a simple premise—instead of defining functions for ad hoc types, we deconstruct our types into a more fundamental type representation and then write generic functions against the lower-level type representation instead.

Instead of writing ad hoc instances of the `foldMap` function for each type, we define a lower-level type representation along with a way to translate all regular datatypes to the lower-level representation. This allows us to write generic functions on the lower level of type representation, impervious to changes in the higher level datatypes.

This is datatype-generic programming—writing generic functions parameterized by the shape of the datatype. It explains why datatype-generic programming is also said to exhibit shape polymorphism, structure polymorphism, or polytypism.

As stated in *Gibbons' Datatype-Generic Programming*:

> *"Parametric polymorphism abstracts from the occurrence of 'integer' in 'lists of integers', whereas datatype genericity abstracts from the occurrence of 'list'"*

There are many patterns of datatype-generic programming that differ in the design of the type representation and the way in which the functions on the type representation are arranged.

For the rest of this chapter, we will explore some key patterns of datatype-generic programming.

The sum of products style

To explore datatype-generic functions in the sum of products style, we'll return to the familiar `List` and `Tree` values:

```
data List' a = Nil' | Cons' a (List' a)
     deriving (Show)

data Tree a = Node a (Tree a) (Tree a)
          | Leaf a
            deriving (Show)

  aList = (Cons' 2 (Cons' 3 (Cons' 5 Nil')))
intTree
 = Node 2 (Leaf 3)
                  (Node 5 (Leaf 7)
                          (Leaf 11))
```

As a reference point, we define the datatype-specific `size` functions:

```
sizeT (Leaf _) = 1
sizeT (Node _ lt rt)
  = 1 + (sizeT lt) + (sizeT rt)

sizeL Nil' = 0
sizeL (Cons' _ xs)
  = 1 + (sizeL xs)
```

As is the case with recursive functions over recursive types, notice how the shape of functions follows the shape of the underlying recursive datatype.

Instead of these ad hoc polymorphic functions, let's write them in a datatype-generic way. First, we define a type representation. In this section, we follow the generic programming style of **Lightweight Implementation of Generics and Dynamics (LIGD)** by Cheney and Hinze, 2002, as revised in *Libraries for Generic Programming in Haskell* by *Jeuring et al.*, 2009.

The sum of products type representation

LIGD uses the sum of products style of type representation. For example, in this style, List' would be deconstructed as either Nil or the combination of an element with another list. In order to represent List' we need to be able to express "Nil", "choice of either", and "combination of":

```
data List' a = Nil' | Cons' a (List' a)
```

Constructors that take no arguments (for example, Nil') are represented as follows:

```
data U = U
   deriving (Show)
```

To simplify this example, we discard the constructor names and focus only on the structure of the datatype. We can encode the choice between multiple constructors in the style of the Either type:

```
data Choice a b = L a | R b
   deriving (Show)
```

that is, List is a choice between U and Cons':

```
Choice U (Cons' a (List' a))
```

To represent the combination of two arguments of Cons', we use the following:

```
data Combo a b = Combo a b
   deriving (Show)
```

Consider the following example:

```
Cons' a (List' a) -> Combo a (List' a)
```

This encodes type constructors with multiple arguments. The List type can now be represented as follows:

```
type RList a = Choice U (Combo a (List' a))
```

Note that the RList function does not recurse but refers to List' instead; this is called a **shallow type representation**.

Similarly, we can represent Tree in this type representation:

```
type RTree a
  = Choice (Combo U a)
              (Combo a
                        (Combo (Tree a)
                                  (Tree a)))
```

Although `Combo` and `Choice` take two arguments, we can express multiple arguments through nesting, for example, in `(Combo a (Combo (Tree a) (Tree a)))`.

In the sum of products representation style, sum refers to our `Choice`, product refers to `Combo`, and unit refers to `U`.

Translating between the type and representation

Now that we have an alternative representation for `List`, we can translate to and from the type representation:

```
fromL :: List' a -> RList a
fromL Nil'          = L U
fromL (Cons' x xs)  = R (Combo x xs)

toL :: RList a -> List' a
toL (L U)         = Nil'
toL (R (Combo x xs))   = (Cons' x xs)

main = do
  print $ fromL aList
  print $ (toL . fromL) aList
```

Before we move on, let's capture the translation functions in one type:

```
data EP d r = EP {from :: (d -> r),
                    to :: (r -> d)}
```

Writing a datatype-generic function

At this stage, our type representation consists of the following three data types:

```
data U = U
data Choice a b = L a | R b
data Combo a b = Combo a b
```

If we are to define a generic function parameterized by this type representation, we'll need to group these disparate types into one. There are different ways of doing this and for our purposes, we'll use a GADT:

```
data TypeRep t where
  RUnit    :: TypeRep U
  RChoice :: TypeRep a -> TypeRep b
          -> TypeRep (Choice a b)
```

```
RCombo  :: TypeRep a -> TypeRep b
          -> TypeRep (Combo a b)
RInt    :: TypeRep Int
Rtype :: EP d r -> TypeRep r -> TypeRep d
```

What makes the `TypeRep` constructor a GADT is that each constructor returns a different specialization of the general type `TypeRep t`. We added two other constructors that will only make sense a bit further down the line: `RInt` and `RType`.

Recall that we defined `Rlist` in terms of the type representation datatypes:

```
type RList a = Choice U (Combo a (List' a))
```

We need a corresponding type based on the `TypeRep` constructors. `rList` creates a more finely-typed list representation that packages the list representation together with `toL` and `fromL`:

```
rList :: TypeRep a -> TypeRep (List' a)
rList tr = RType (EP fromL toL)
                 (RChoice RUnit
                 (RCombo tr (rList tr)))
```

The `rList` is a recursive function using the `TypeRep` constructors as building blocks. The first argument (`TypeRep a`) guides the type resolution of `List'`; for example, if we passed in (`TypeRep Int`), the resulting type would be:

```
TypeRep Int -> TypeRep (List' Int)
```

This is where we need the `RInt` function:

```
rList (TypeRep Int)  -- INVALID
rList RInt           -- VALID
```

Similarly, we would need additional constructors for `RFloat`, `RDouble`, `RChar`, and so on, to deal with lists of these types.

Now, we can write a generic function, say `gSize`, parameterized by both the type representation and the instance of the type:

```
gSize :: TypeRep a -> a -> Int
gSize RUnit  U  = 0
gSize (RChoice trA trB) (L a)
  = gSize trA a
gSize (RChoice trA trB) (R b)
  = gSize trB b
gSize (RCombo  trA trB) (Combo a b)
```

```
      = (gSize trA a) + (gSize trB b)
  gSize RInt   _ = 1
  gSize (RType ep tr) t
    = gSize tr (from ep t)
```

GADTs give a finer precision of types in data constructors, and hence a finer precision when we perform pattern matching against algebraic datatypes.

Finally, we can apply the `gSize` function to (`List' Int`):

```
{-# LANGUAGE ExistentialQuantification #-}
{-# LANGUAGE GADTs #-}


main = print $ gSize (rList RInt) aList
```

It's worth evaluating this in slow motion:

```
    gSize (rList RInt) aList
    gSize (RType ep listRep) aList
    gSize listRep (from ep aList)

  -- substitute listRep, apply 'from' to aList
  gSize (RChoice RUnit (RCombo RInt (rList RInt)))
        R (Combo 2 (Cons' 3 (Cons' 5 Nil')))

  -- choose the 2nd type-rep because of R in list rep
  gSize (RCombo RInt (rList RInt))
        (Combo  2    (Cons' 3 (Cons' 5 Nil')))

  -- add the matching type-rep and list-rep pairs
  (gSize RInt 2)
        + (gSize (rList RInt) (Cons' 3 (Cons' 5 Nil')))

  -- evalulate (gSize RInt 2)
  1   + (gSize (rList RInt) (Cons' 3 (Cons' 5 Nil')))
  ...
```

Note how the `TypeRep` constructors guide the pattern matching of the appropriate data contents.

Adding a new datatype

Adding a new datatype, say `Tree`, does not require amending the generic functions. (Of course, we may want to override a generic function for a particular type. Generic libraries have different ways to allow this, but we'll omit those details in this investigation).

We already defined `RTree` and are left with `fromT`, `toT`, and `rTree`:

```
fromT :: Tree a -> RTree a
fromT (Leaf x)       = L (Combo U x)
fromT (Node x lt rt) = R (Combo x (Combo lt rt))

toT :: RTree a -> Tree a
toT (L (Combo U x))
  = Leaf x
toT (R (Combo x (Combo lt rt)))
  = (Node x lt rt)

rTree :: TypeRep a -> TypeRep (Tree a)
rTree tr
  = RType (EP fromT toT)
              (RChoice (RCombo RUnit tr)
                       (RCombo tr
                               (RCombo (rTree tr)
                                       (rTree tr))))
```

Now, we use the datatype-generic `gSize` function on `Tree`:

```
main = print $ gSize (rTree RInt) intTree
```

This may seem like a lot of work to gain one generic function for `Tree`, but we can now use the whole class of generic functions defined against the underlying type representation. If we were to add `gEq`, `gShow`, `gFold`, `gTraverse`, and so on, they would all be automatically applicable to `Tree` and `List`.

GHC.Generics – a generic deriving mechanism

In 2010, a more generic approach was introduced to synthesize (and surpass) the Derivable type-classes of Haskell 98 (refer to *A Generic Deriving Mechanism for Haskell* by *Magalhaes et al*, 2010)

The generic deriving mechanism was based on a more sophisticated extension of the sum of products approach of this section. Furthermore, the new approach also enabled autoderiving of user-defined type classes.

This work has found its way into the Haskell ecosystem in the form of the GHC.Generics library.

Origami programming

> *"Recursive equations are the 'assembly language' of functional programming, and direct recursion the go-to"*
>
> *-- Jeremy Gibbons, Origami Programming (The Fun of Programming), 2003*

In the previous section, we wrote a generic function for the recursive types `Tree` and `List`. In this section, we look at origami programming, a style of generic programming that focuses on the core patterns of recursion: `map`, `fold`, and `unfold`.

Tying the recursive knot

There is a primal type that underlies the recursive datatypes, known as `Fix`:

```
data List' a = Nil'   | Cons' a (List' a)
data Tree  a = Leaf a | Node  a (Tree a) (Tree a)

data Fix s a = FixT {getFix :: s a (Fix s a)}
```

The parameter s represents the shape, while a refers to an instance of the type. A `Fix` datatype is named after a fixed point of a function, which is defined by the following:

```
f (fix f) = fix f
```

To express `Tree` and `List` in terms of `Fix`, we need to rewrite them with an implicit recursion:

```
data List_ a r = Nil_     | Cons_ a r
  deriving (Show)
data Tree_ a r = Leaf_ a | Node_ a r r
  deriving (Show)
```

Here, we replaced the explicit recursive reference with a more vague type parameter r. This allows us to express ListF and TreeF in terms of Fix:

```
type ListF a = Fix List_ a
type TreeF a = Fix Tree_ a
```

The `List_` and `Tree_` functions don't explicitly recur, so `Fix` ties the recursive knot around the shape (refer to *Datatype-Generic Programming* by *Gibbons*). We can construct a `List_` function in a similar way:

```
aList1 = Cons_ 12 Nil_
-- aList :: List_ Integer
          (List_ a r)


aList2 = Cons_ 12 (Cons_ 13 Nil_)
-- aList2 :: List_ Integer
        (List_ Integer (List_ a r))
```

Although the type signatures show that this is quite different from regular lists, to construct the `ListF` lists, we need to wrap the `FixT` constructor around each nesting:

```
aListF :: ListF Integer -- Fix List_ Integer
aListF
   = FixT (Cons_ 12
                            (FixT (Cons_ 13
                                        (FixT Nil_))))
```

The generic map

We deconstructed `List` and `Tree` into the generic recursion type `Fix`. This means we can write generic functions against the `Fix` type.

To find our way toward a generic map, let's write down a map for the fixed recursive type `ListF`:

```
mapL f listF = case list_ of
    (Cons_ x r)
          -> FixT $ Cons_ (f x) (mapL f r)
    Nil_
          -> FixT Nil_
   where  list_ = getFix listF

showListF :: (Show a) => ListF a -> String
showListF (FixT (Cons_ x r))
  = (show x) ++ ", " ++ (showListF r)
showListF (FixT (Nil_)
  = "Nil_"

main = putStrLn . showListF $ mapL (*2) aListF
```

This is clumsy because we have to unwrap the list (with `getFix`) and then rewrap the result with `FixT` in both base clauses. We need to find a better abstraction for `mapL`, as stated in *Datatype-Generic Programming* by *Gibbons*:

> *"It turns out that the class Bifunctor provides sufficient flexibility to capture a wide variety of recursion patterns as datatype-generic programs."*

The `Bifunctor` class is just a `Functor` class that can have two functions applied to it instead of one (which is already defined in `Data.Bifunctor`):

```
class Bifunctor s where
   bimap :: (a -> c) -> (b -> d) -> (s a b -> s c d)
```

Let's make the `List_` and `Tree_` instances of the `BiFunctor` class:

```
instance Bifunctor List_ where
  bimap f g Nil_        = Nil_
  bimap f g (Cons_ x r) = Cons_ (f x) (g r)

instance Bifunctor Tree_ where
  bimap f g (Leaf_ x)        = Leaf_ (f x)
  bimap f g (Node_ x rl rr) = Node_ (f x) (g rl) (g rr)
```

Now, we can write a generic map:

```
gmap :: Bifunctor s => (a -> b) -> Fix s a -> Fix s b
gmap f = FixT . bimap f (gmap f) . getFix


main = putStrLn . showListF $ gmap (*2) aListF
```

The generic fold

The `bimap` function also gives us a generic fold:

```
gfold :: Bifunctor s => (s a b -> b) -> Fix s a -> b
gfold f = f . bimap id (gfold f) . getFix
```

Again, we unwrap the list with `getFix`, but this time, instead of rewrapping, we apply `f`. In other words, `gfold` replaces the occurrences of `FixT` with `f`:

```
-- FixT (Cons_ 12 (FixT (Cons_ 13 (FixT Nil_))))
-- f    (Cons_ 12 (f    (Cons_ 13 (f    Nil_))))
```

To fold together a sum, we create an adder:

```
addL (Cons_ x r) = x + r
addL Nil_        = 0
```

main = print $ gfold addL aListF

Where `fold` is a consumer of data structures, `unfold` is a producer that unfolds a structure from a single value. To unfold a regular list, we need a value and some functions:

```
unfoldL stopF nextF val
  = if stopF val -- stop if True
    then []
    else val :
         (unfoldL stopF nextF (nextF val))

main = print $ unfoldL (< (-10)) (\x -> x - 1) 10
```

We can use the `bimap` function to create a generic unfold:

```
gunfold :: Bifunctor s => (b -> s a b) -> b -> Fix s a
gunfold f = FixT . bimap id (gunfold f) . f
```

Consider the following example:

```
toList 0 = Nil_
toList n = (Cons_ n (n-1))

main = putStrLn . showListF $ gunfold toList 10
--       10, 9, 8, 7, 6, 5, 4, 3, 2, 1, Nil_
```

Generic unfold and fold

Composing unfold with fold makes sense because by doing so, we are connecting a producer with a consumer:

```
main = print $ gfold addL (gunfold toList 100)
```

(Refer to *Gibbons' origami programming* and *datatype-generic programming*, where you will find them under *hylomorphisms*.)

The `unfold` and `fold` functions are each other's mirror images:

```
gunfold f = FixT . bimap id (gunfold f) . f
gfold f   = f    . bimap id (gfold f)   . getFix
```

The `hylo` function is their composition:

```
hylo f g = g . bimap id (hylo f g) . f
main = print $ hylo toList addL 100
```

> *"The beauty of all of these patterns of computation is the direct relationship between their shape and that of the data they manipulate"*

> *"The term origami programming for this approach because of its dependence on folds and unfolds"*

-- Gibbons, Datatype-Generic Programming

Origami design patterns

Jeremy Gibbons demonstrates that the four key Gang of Four design patterns are captured by origami recursion operators:

- **Composite pattern**: This is the center of the origami pattern constellation. Recursive datatypes express the composite design pattern (as we discussed in *Chapter 1, Functional Patterns – the Building Blocks*).

  Since the `Fix` recursive captures a whole class of recursive datatypes, `Fix` captures the composite pattern most concisely.

- **Iterator pattern**: An iterator gives linear access to the parts of a composite pattern in such a way that the shape of the structure being traversed is preserved.

  In *Chapter 4, Patterns of Folding and Traversing*, we discussed how the iterator pattern is captured most generally by applicative traversals.

  We can construct generic applicative traversals in the origami style we have seen here, and thereby capture the iterator pattern.

- **Visitor pattern**: In *Chapter 1, Functional Patterns – the Building Blocks*, we saw how this pattern is captured by polymorphic dispatch. More generally, the visitor pattern is concerned with structured traversal of a composite; this is precisely what `fold` enables.

  The generic fold allows to visit the elements of a composite pattern. Visitor, like `fold`, does not generally preserve the shape of the data structure being visited.

  This is why it is said that the iterator views that composite as a "container of elements", and the visitor views the composite shape as insignificant.

- **Builder pattern**: This pattern separates the construction of a complex object from its representation, so that the same construction process can create different representations (refer to *Design Patterns* by *Gamma et al.*).

We can express the structured construction of data structures with unfold. The hybrid `hylo` function (`unfold` + `fold`) also allows for data generation. In this way the origami style captures the builder pattern.

 Refer to *Jeremy Gibbons' Design Patterns as Higher-Order Datatype-Generic Programs* for more information on this.

Scrap your boilerplate

Scrap your boilerplate (**SYB**) is another early approach to datatype-generic programming, that is, it provides a way to define generic functions over a "universal" type representation.

SYB differs from the other two approaches we explored in that the type representation is obfuscated from the user.

Earlier, the SYB approach had a strong focus on generic traversals, as we will see in the following example, where we create a generic traversal over a complex nested data structure:

```
data Book     = Book     Title [Chapter]
data Chapter  = Chapter Title [Section]
data Section  = Section Title [Para]
type Title    = String; type Para = String

haskellDP = Book "Haskell Design Patterns" chapters

chapters  = [Chapter "Chapter 1" sections1,
    Chapter "Chapter 2" sections2]

sections1 = [Section "1.1" ["S1"],
    Section "1.2" ["S1.2.1", "S1.2.2"]]

sections2 = [Section "2.1" ["S2.1"],
    Section "2.2" ["S2.2.1", "S2.2.2"]]
```

Suppose we have a function, fSection, which we want to apply to all sections embedded in a Book. This would be a great job for the Lens library, which can precisely deliver a function to elements in a complex structure, but we will follow a different route here:

```
fSection (Section t lines') = Section "!!!" lines'

main = print $ fSection (Section "1.1" ["S1"])
```

Our strategy will be to morph our function into one that can be applied to all parts of the data structure (Book). The new function should simply ignore elements for which it was not intended.

The type-safe cast with typeable

To make type-safe cast possible, we will have to do a type comparison to check whether an element of a book is in fact a section.

The Data.Typeable type-class has what we need in the form of the type-safe cast function. We can autoderive Typeable for our types (in fact, since GHC 7.8, the compiler does not allow us to implement Typeable ourselves). For this, we need to add a language pragma for the DeriveDataTypeable function:

```
{-# LANGUAGE DeriveDataTypeable #-}
import Data.Typeable

data Book     = Book    Title [Chapter]
   deriving(Show, Typeable)
data Chapter  = Chapter Title [Section]
   deriving(Show, Typeable)
data Section  = Section Title [Para]
   deriving(Show, Typeable)

type Title = String; type Para = String
```

Let's get to know the Typeable.cast function:

```
ghci> cast 'a' :: Maybe Char
      -> Just 'a'
ghci> cast 'a' :: Maybe Int
      -> Nothing
```

If the value type matches with the type inside `Maybe`, we get the `Just` value, otherwise we get `Nothing`:

```
ghci> (cast (Section "t" ["s1", "s2"])) :: Maybe Section
    -> (Just (Section "title" ["line1", "line2"]))

ghci> (cast (Book "title" [])) :: Maybe Section
    -> Nothing

Since functions are also just types, we can cast them too!

ghci> cast (++ "a") :: Maybe String
    -> Nothing
ghci> cast (++ "a") :: Maybe (String -> String)
    -> Just (++ "a")
```

Type-safe function application

Let's write a higher order function to enable the type-safe function application:

```
typesafeF :: (Typeable a, Typeable b) =>
    (b -> b) -> a -> a

typesafeF f
  = case (cast f) of
        Just f' -> f'
        Nothing -> id
```

Consider the following example:

```
ghci> typesafeF (+1) 3
    -> 4
ghci> typesafeF (+1) "3"
    -> "3"
```

By lifting `fSection` into a type-safe function, we can apply it to any part of the `Book` type (or any `Typeable` type, for that matter):

```
main = do
  print $ (typesafeF fSection) aSection
  print $ (typesafeF fSection) aBook
  where
    aSection = (Section "1.1" ["s1", "s2"])
    aBook = (Book "title" [])
```

The (typesafeF fSection) leaves all values not targeted by the fSection function, that is, in the first call in the preceding code, typesafeF will apply fSection to Section. In the second call, typesafeF applies id to Book.

The shallow traversal and the data type-class

In both the sum of products and origami styles of generic programming, we encountered shallow recursion in crucial parts of the formulation.

Let's explore how we might traverse a Book class with a function, again favoring a shallow approach. Starting with traversal of a Chapter variable, we will write the following:

```
gmap f (Chapter title sections)
  = Chapter (f title) (f sections)
-- INVALID
```

We quickly run into trouble! The function cannot span all the different types it is being applied to. This is why we need to resort to Rank2Types so that we can define a gmap function that accepts a generic mapping function. (We will delve more deeply into higher rank types in *Chapter 5, Patterns of Type Abstraction*).

We also make gmap a part of the Data' type-class to be implemented by all the types we plan to traverse over:

```
{-# LANGUAGE Rank2Types #-}

-- Data' inherits from Typeable
class Typeable a => Data' a where
  gmap :: (forall b. Data' b => b -> b) -> a -> a

-- note the shallow recursion in gmap implementations...

instance Data' Book where
  gmap f (Book title chapters)
    = Book (f title) (f chapters)

instance Data' Chapter where
  gmap f (Chapter title sections)
    = Chapter (f title) (f sections)

instance Data' Section where
  gmap f (Section title paras)
```

```
      = Section (f title) (f paras)

  instance Data' a => Data' [a] where
    gmap f []     = []
    gmap f (x:xs) = f x : f xs

  instance Data' Char where
    gmap f c = c

main = do
    print $ gmap (typesafeF fSection) chapter
    print $ gmap (typesafeF fSection) sections1
    where chapter = (Chapter "The building blocks" sections1)
```

If we traverse a chapter with the fSection function, the sections are not reached by the traversal. When we traverse a list of sections, only the first is affected by fSection, while the subsequent sections are ignored by the fSection function:

```
gmap f (section1: sections)
--    = f section1 : f sections
--      f sections = f sections
--                 = f [Section]
--                 = id [Section]
```

This shallow traversal of the gmap function can seem a bit pointless, but the big advantage it brings is that we can mold it into different kinds of recursions, for example:

```
-- bottom-up traversal: traverse x before applying f
traverse :: Data' a =>(forall b . Data' b => b->b) -> a -> a
traverse f x = f (gmap (traverse f) x)


-- vs top-down traversal: apply f x then traverse

traverse' :: Data' a =>(forall b . Data' b => b->b) -> a -> a
traverse' f x = gmap (traverse' f) (f x)

main = do
    -- this time our traversal is reaching all Sections...
    print $ traverse (typesafeF fSection) chapter
    print $ traverse (typesafeF fSection) sections1
    print $ traverse' (typesafeF fBook) haskellDP
    where
      chapter = (Chapter "The building blocks" sections1)
      fBook (Book t chapters) = Book "!!!" chapters
```

Together, the `gmap` and `typesafeF` functions enable us to deliver type-specific functions deeply into nested structures.

> *"Recursive traversal in two steps — first define a one-layer map, and then tie the recursive knot separately — is well-known folk lore in the functional programming community. For lack of better-established terminology we call it 'the non-recursive map trick'"*
>
> *-- Lammel and Peyton Jones, 2003, Scrap your Boilerplate: A Practical Design Pattern for Generic Programming*

Typeable and data

Our `data'` type-class mimics the Haskell `Data` type-class, which is based on the more general `gfoldl` (as opposed to our `gmap`).

The SYB approach relies on the compiler to auto-generate instances of `Typeable` and `Data`. This is a purposeful obfuscation of the underlying structure of datatypes and contrasts with the sum of products and origami styles, which involve the programmer directly in the translation to and from the relevant type representation.

In SYB, the programming interface consists of generic combinators based on `Typeable` and `Data`. This is why it is said that `Typeable` provides the backend of SYB and the data provides the frontend.

Even though we are not directly engaged with the type representation, we are still dealing with the generic functions defined in the abstract data shape, where `Data` provides the abstraction over shapes made of `Typeable`'s.

Scrap your boilerplate

We have just scratched the surface of this technique. In the above referenced paper, Lammel and Peyton Jones cover the following points:

- Generalizing to traversals that transform the shape of a data structure ("queries")
- Generalizing the monadic traversals
- Unifying all of the preceding techniques with a generic `fold`

In further work, SYB was extended to allow for extensibility, in the sense of being able to override generic behavior for a given type.

SYB was an early approach that strongly influenced the development of datatype-generic programming in Haskell.

For example, the `Uniplate` library represents a simplified (and less powerful) phrasing of SYB. The `Uniplate` library has since been embedded in the `Lens` library.

Summary

This chapter started with a high-level discussion of the patterns of generic programming.

In the remainder of the chapter, we explored three variations of datatype-generic programming, that is, three techniques of parameterizing generic functions by data shape rather than by contents.

Underlying these varied approaches were some common threads:

- To achieve generic functions, we need "generic data" (in the form of a universal type representation)
- Datatype-generic programming mixes well with more rudimentary meta-programming (for example, when we autoderived `Typeable` instances in the process of writing generic functions)
- using "shallow recursion" on a lower layer allows us to express all manner of recursive functions on the higher level

Generic Programming relies on extensions to the Haskell Type system. In the following *Chapter 7, Patterns of Kind Abstraction* we will see that we can raise the level of abstraction even further via extensions to the Kind system!

7

Patterns of Kind Abstraction

Type-level programming involves computation at type-level that is executed during the type-checking phase.

In earlier chapters, we saw the beginnings of type-level programming when we used functional dependencies and GADTs, but to do proper type-level programming, we need an even more enriched kind system.

We will freely interchange the terms type-level and kind-level, since in Haskell, type-level programming happens at the level of kinds.

Type-level programming stands in contrast to term-level programming, in that the former executes in the type-checking phase and the latter executes at runtime.

In this chapter, we will explore patterns of kind-abstraction as they relate to type-level programming. First, we will look at the basic kind system, extend it with the associated types, and then look at more generalized type families.

In the last round of extensions, we will add kind polymorphism and datatypes (via data type promotion) to the kind system. Together, these extensions form the basis for type-level programming in Haskell.

We finish the chapter and the book at the very edge of Haskell when we look into Dependently-typed programming.

In this chapter, we will cover the following topics:

- Higher-order kinds
- Higher-kinded polymorphism
- Associated type synonyms
- Type (synonym) families
- Kind polymorphism

- Type promotion
- Type-level programming
- Dependently-typed programming

Higher-order kinds

Types classify values at different levels of abstraction, for example:

```
-- Primitive types
"a string"  :: String
12              :: Int

-- instances of higher-order, parameterized types
Just 10  :: Maybe Int
Left 10  :: Either Int b

-- functions are first class values
(* 2) :: Num a => a -> a

-- and type-constructors are functions
Just :: a -> Maybe a
Left :: a -> Either a b
```

In a similar way, kinds classify types. For monomorphic types (that is, not polymorphic), the kind signature is just the placeholder *:

```
-- TYPE          KIND
[Char]          ::     *
Maybe Int    ::     *
```

Parametric types express higher-order kinds, for example:

```
Maybe  ::  * -> *
--            a -> Maybe a
```

where (* -> *) is a placeholder for a -> Maybe a. Let's compare this with the kind signature of Either:

```
Either        * -> *  -> *
--                a -> b -> Either a b
```

The Haskell kind system can distinguish between two main kinds: lifted types (of kind *) and type constructors (for example, * -> * -> *).

Kind signatures signify the arity of parameterization of a type, that is, the number and position of type parameters in a type. However, arity says nothing about type; the Haskell kind system is untyped.

Higher-kinded polymorphism

Type-classes with one parameter type have kind of (* → *) operator, for example...:

```
class Show  a :: * -> *
class Maybe a :: * -> *
```

If we declare an instance of Show, as shown in the following code:

```
instance (Show a) => Show (Maybe a) where ...
```

Then the type-class parameters in the kind signatures need to be aligned, we use this. In order to match the kind of a :: * in Show a, we use Maybe' b instead of Maybe:

```
Maybe' b :: *
-- instead of
Maybe :: (* -> *)
```

The Monad type-class is of a higher-order than Show and Maybe:

```
class Monad m :: (* -> *) -> *
--                            m      -> Monad m
```

The Show type-class is parameterized over type a :: *; whereas Monad is parameterized over the type constructor m :: * -> *. This can seem like a natural and unsurprising generalization for type-classes, but was in fact an exciting leap forward for Haskell, as described in *Hudak et al's History of Haskell*:

> *"The first major, unanticipated development in the type-class story came when Mark Jones suggested parameterizing a class over a type constructor instead of over a type...*

> *"Jones's paper appeared in 1993, the same year that monads became popular for I/O. The fact that type classes so directly supported monads made monads far more accessible and popular; and dually, the usefulness of monadic I/O ensured the adoption of higher-kinded polymorphism."*

Associated type synonyms

As a unifying example for the next sections, we return to *Chapter 6: Patterns of Generic Programming*, where we created a `List'` object and its type representation `RList`:

```
data List' a
  = Nil' | Cons' a (List' a)
  deriving (Show)

data U = U
  deriving (Show)

data Choice a b = L a | R b
  deriving (Show)

data Combo a b = Combo a b
  deriving (Show)

type RList a = Choice U (Combo a (List' a))
```

In addition to this, we defined functions `fromL` and `toL` to mediate between the type and representation:

```
fromL :: List' a -> RList a
toL   :: RList a -> List' a
```

We embedded this in the container type `EP` as follows:

```
data EP d r = EP {from_ :: (d -> r),
                        to_ ::  (r -> d)}
```

Using functional dependencies

Instead of the container type EP mentioned above, we could have used a multi-parameter type-class with functional dependencies:

```
-- requires
-- {-# LANGUAGE FlexibleInstances #-}
-- {-# LANGUAGE FunctionalDependencies #-}
class GenericFD d r | d -> r where
  from :: d -> r
  to   :: r -> d

instance GenericFD (List' a) (RList a) where
  from Nil'          = L U
  from (Cons' x xs)  = R (Combo x xs)
  to (L U)           = Nil'
```

```
    to (R (Combo x xs))  = (Cons' x xs)

    from :: GenericFD d r => d -> r
    from (Cons' "1" Nil') :: RList [Char]
    from (Cons' 1 Nil') :: Num a => RList a

    main = print $ from (Cons' "1" Nil')
```

The `from` function is constrained to functionally-related types `d` and `r`. The functional dependency (`d -> r`) tells the compiler to only accept one `r` for every `d`; for example, we can't declare an alternative target representation for (`List' a`):

```
    instance GenericFD (List' a) (AltRList a) ...
```

Multiparameter type-classes became more useful once there was a way to constrain the relationship between the parameters (in the absence of which, type inference is not possible).

Functional Dependencies by *Mark Jones*, 2000, was the first solution to this problem. They introduced the notion of type function, albeit implicitly, through relations. Type functions, in turn, unleashed a wave of type-level programming in the Haskell community.

In 2002, five years after the introduction of functional dependencies, associated type synonyms were introduced as an alternative way to specify a relationship between multiple type-class parameters as explicit type functions.

Let's rewrite our `Generic` type-class using the associated types.

Associated type synonyms

The key observation that leads us from functional dependencies to associated type synonyms is that the (`GenericFD d r`) type-class doesn't really have two parameters, but rather one parameter `d`, which uniquely determines the other parameter `r`:

```
    -- {-# LANGUAGE TypeFamilies #-}
    class GenericA d where
      type Rep d :: *

      fromA :: d        -> (Rep d)
      toA   :: (Rep d)  -> d
```

The `Rep` is a type function (or type family, or associated type). In contrast to functional dependencies, the associated type synonym makes the `type` function explicit.

The `fromA` and `toA` are generic functions that are indexed against types that are themselves indexed by types! In this way, associated type synonyms extend type-classes by allowing for type-indexed behavior.

The type-class instance needs to specify a value for the type function `Rep`, that is, the instance mixes type functions with type-class functions.

```
instance GenericA (List' a) where
  type Rep (List' a) = (RList a)
  -- Rep type params must match the class params

 fromA Nil'              = L U
 fromA (Cons' x xs)    = R (Combo x xs)
 toA (L U)               = Nil'
 toA (R (Combo x xs)) = (Cons' x xs)

 main = print $ fromA (Cons' 1 Nil')
```

This is precisely how generics are implemented in the GHC (`https://wiki.haskell.org/GHC.Generics`). Moreover, `GHC.Generics` provides automatic instance generation with `deriving Generic`.

Associated types versus functional dependencies

It turns out that associated types and functional dependencies have similar expressive power. Having said that, associated types have some clear benefits:

- Associated types provide explicit type functions contrary to the implicit relations of functional dependencies
- Type functions allow us to reduce the number of type parameters
- Type functions are more idiomatically functional than relational-style functional dependencies

Type (synonym) families

In 2008, three years after the introduction of associated types, they were subsumed within the broader framework of type families. Associated types are special type families where the type function is attached to a type-class.

In contrast to associated types, we have top-level type-families that are not associated to a type-class, for example:

```
type family RepF d
type instance RepF (List' a) = (RList a)
```

The type family `RepF` represents a type function, with each instance declaring a value. Put another way, a type family represents a set of types, and each instance represents a set member.

In our example, `GenericF` simply uses the top-level type function in its type signatures, as shown in the following code:

```
class GenericF d where
  fromF :: d          -> (RepF d)
  toF   :: (RepF d)   -> d

instance GenericF (List' a) where
  fromF Nil'           = L U
  fromF (Cons' x xs)   = R (Combo x xs)
  toF (L U)        = Nil'
  toF (R (Combo x xs)) = (Cons' x xs)

main = print $ fromF (Cons' 1 Nil')
```

With associated types, we need to align the type function parameters with the type-class parameters. Top-level type families don't have that restriction and therefore are more general than associated types. The fundamental difference is in the scope of the type function, which boils down to purely a design decision.

Type families are to regular types what type class methods are to regular functions. Instead of polymorphism over value, type families give us polymorphism over datatypes.

As with type-classes, type families are open in the sense that we can add new instances at any time.

One key application of type families is demonstrated in our example: Generic programming. Another is in writing highly-parameterized libraries.

Data families

In our discussion of type families, associated to a type-class or top-level, we used type-synonym families. However, type families can be defined for datatypes as well. This lets us create families of datatypes. As with type-synonym families, datatype families can be associated to the type-class or top-level:

```
-- associated data family
class GMap k where
  data GMap k :: * -> *
  empty      :: GMap k v
  lookup     :: k -> GMap k v -> Maybe v
  insert     :: k -> v -> GMap k v -> GMap k v

-- top-level data family
data family GMap k :: * -> *
-- ...
```

The Gmap function describes a generic interface for associative maps where the type of key k determines the type of the associated value (via the Gmap type function).

Kind polymorphism

Type families bring functions to the type-level through type functions. Similarly, the Polykinds language extension brings polymorphism to kinds, for example, to the type-level.

Kind polymorphism (*Giving Haskell a Promotion,* by *Yorgey et al* in 2012) allows us to describe more generic data and functions. For example, when designing a type-class, the need may arise to cater for various kind-orders. Consider the multiple Typeable classes for multiple arities as an example:

```
class Typeable (a :: * ) where
  typeOf :: a -> TypeRep

class Typeable1 (a :: * -> *) where
  typeOf1 :: forall b. a b -> TypeRep
```

The same goes for the Generic type-class we encountered earlier: we need to define different type-classes for different kind arities.

To explore kind polymorphism, we'll work with a trivialized version of `Typeable`, where the `typeOf` function simply returns a string instead of a `TypeRep` function:

```
class T0 a where
   f0 :: a -> String

instance T0 Int where
   f0 _ = "T0 Int"

instance T0 Char where
   f0 _ = "T0 Char"
```

The class function `f0` gives us the type as a string for each `Int` or `Char` value:

```
-- f0 (10::Int)
--    "T0 Int"
-- f0 'x'
--    "T0 Char"
```

We can involve higher-kinded types as instances, such as `Maybe`:

```
instance T0 (Maybe a) where
   f0 _ = "T0 Maybe a"
```

However, we have to specify parameter a for `Maybe :: * -> *` to match the required kind `*` of `T0`. in order to achieve "instance T0 Maybe", we need to create an alternative type-class to deal with the higher-kinded case:

```
class T1 m where -- m :: * -> *
   f1 :: Show a => m a -> String

instance T1 Maybe where
   f1 _ = "T1 Maybe"
```

We can deal with the higher-kinded case but at the expense of the monomorphic case:

```
-- f1 (Just 10)
--    "T1 Maybe"

-- but not
--   instance T1 Int where
--      f1 _ = "T1 Int"
```

Polymorphic kinds allow us to unify these type-classes into one.

The PolyKinds language extension

A kind-polymorphic type-class looks quite normal at the first glance:

```
-- {-# LANGUAGE PolyKinds #-}
class T a where -- (a::k)
  f :: Proxy a -> String
```

Type variable `a` has kind variable `k`. With the `PolyKinds` language extension, `k` is polymorphic by default, that is, `k` can take several forms (of kind signatures):

```
class T a where
-- (a::k)

-- *
-- * -> *
-- * -> * -> *
```

The `k` variable is a polymorphic placeholder for many possible kind arities.

The `Proxy` variable is a kind-polymorphic phantom-type:

```
data Proxy a = Proxy -- (a::k)
  deriving Show
```

The type variable 'a' has polymorphic kind, for example:

```
(Proxy Int)     -- Proxy :: *          -> *
(Proxy Maybe)   -- Proxy :: (* -> *) -> *
```

The `Proxy` variable is used to generalize the kind of function argument, for example:

```
f :: T a => Proxy a -> String -- types
--     k  =>    k     -> *       -- kinds
```

The first argument of `f` can take a type with any kind-order (arity). Note that the type `a` in `Proxy a` is constrained by the type-class `T a`, which is also kind polymorphic in the type `a`.

If we interrogate the kinds before and after the `PolyKinds` language extension (for example, `ghci> :k f` gives the kind signature of `f`), we can see how the kind signatures are generalized by `PolyKinds`:

```
-- kind signatures: before and after PolyKinds

-- before
f  :: *            -> *
-- after
```

```
f  ::  forall k. k      -> *

-- before
Proxy :: *              -> *
-- after
Proxy :: forall k. k -> *
```

Note how the first argument of f is kind-polymorphic (thanks to the kind-polymorphic phantom type Proxy).

The forall keyword makes an appearance on the type-level in a way similar to how it is used with the RankN types to describe nested function polymorphism. We can verify that the type parameter has polymorphic kind:

```
instance T Int where -- Int :: *
  f _ = "T Int"

instance T Maybe where -- Maybe :: * -> *
  f _ = "T Maybe"

-- f (Proxy :: Proxy Maybe) -- "T Maybe"
-- f (Proxy :: Proxy Int)   -- "T Int"
```

The particular type of Proxy we pass to f determines the type-class at which the function will be invoked. In this example, kind-polymorphism appears in three different guises:

- The T is a kind-polymorphic type-class
- The Proxy datatype is kind-polymorphic (it takes types of any kind-order and returns a * kind)
- The Proxy data-constructor is a kind-polymorphic function

Regular polymorphism over functions and types increases the opportunities for abstraction. The same is true for kind polymorphism at the type level.

Type promotion

Type promotion was introduced in the same paper as kind polymorphism (*Giving Haskell a Promotion*, by *Yorgey et al* in 2012). This represented a major leap forward in Haskell's type-level programming capabilities.

Let's explore type promotion in the context of a type-level programming example. We want to create a list where the type itself contains information about the list size.

To represent numbers at type level, we use the age-old Piano numbering which describes the natural numbers (1,2,3, ...) in a recursive manner:

```
data Zero = Zero
  deriving Show
data Succ n = Succ n
  deriving Show


one = Succ Zero
two = Succ one
```

We'll use this with the understanding that certain bad expressions are still allowed:

```
badSucc1 = Succ 10    -- :: Succ Int
badSucc2 = Succ False -- :: Succ Bool
```

Our size-aware list type `Vec` is represented as a GADT:

```
-- requires
-- {-# LANGUAGE GADTs #-}
-- {-# LANGUAGE KindSignatures #-}

data Vec :: (* -> * -> *) where
  Nil  :: Vec a Zero
  Cons :: a -> Vec a n -> Vec a (Succ n)

nil' = Nil :: Vec Int Zero

cons1 = Cons 3 nil'
--         :: Vec Int (Succ Zero)

cons2 = Cons 5 cons1
--         :: Vec Int (Succ (Succ Zero))
```

The `Vec` data type has two type parameters: the first represents the list data and the second represents the list size, that is, either `Zero` or `Succ` (although this is not enforced by the type-checker).

The `Cons` function increments the list size as part of the type signature. Every time we cons an element to a list, a different list type is returned, and thereby the list size is incremented. Instead of a size function defined on term-level, we now have a type-level function.

Unfortunately the following is valid...:

```
badVec = Nil :: Vec Zero Zero
```

However, the following is not valid:

```
-- badVec2 = Nil :: Vec Zero Bool -- INVALID
```

We need more type-safety when working with kinds, and for that, we need datatypes on the kind level. That is precisely what the `DataKinds` extension enables.

Promoting types to kinds

The `Vec` data type expresses type-level programming, but with a very blunt tool, where kinds only describe arity of types and a little more.

The `DataKinds` language extension promotes all (suitable) datatypes to kind level in such a way that we can use the types as kinds in kind signatures. This gives us type-safety at the type-level. Let's continue with the example from the previous section. First, we unify `Zero` and `Succ` into the `Nat` datatype, as follows:

```
-- {-# LANGUAGE DataKinds #-}
data Nat = ZeroD | SuccD Nat
```

This gives us more type-safety:

```
badSuccD = SuccD 10 -- INVALID
```

The `DataKinds` language extension will automatically promote the `Nat` type to the `Nat` kind. The data-constructors `ZeroD` and `SuccD` are promoted to types.

The revised `Vec` datatype uses type promotion along with kind polymorphism. The second type parameter is now constrained to be of the `Nat`:

```
data VecD :: * -> Nat -> * where
  NilD :: VecD a 'ZeroD
  ConsD :: a -> VecD a n -> VecD a ('SuccD n)

cons1D = ConsD 3 NilD
--       :: VecD Integer ('SuccD 'ZeroD)

cons2D = ConsD '5' NilD
--       :: VecD Char ('SuccD 'ZeroD)
```

Promoted types and kinds can be prefixed with a quote `'` to unambiguously specify the promoted type or kind.

The type signature `ConsD` uses type `'SuccD` (promoted from the datatype constructor `SuccD`). Similarly, `NilD` uses the type `'ZeroD`.

We created custom kinds by promoting custom datatypes, then we used them to express constrained kind signatures.

Type-level programming

Type families bring functions to the type-level. Polymorphic kinds bring polymorphism to the kind-level. Type promotion bring datatypes and type-safety to the kind-level.

Two major problems of the Haskell kind system are solved by these extensions:

- The kind system is too restrictive (because it lacks polymorphism)
    - ° **Solution**: Provide polymorphism on the kind level (PolyKinds)
- The kind system is too permissive (kinds are too vague)
    - ° **Solution**: Promote datatypes to kinds to simulate a type-system on the kind-level (DataKinds)

Haskell98 already carried the seed for type-level programming by including multiparameter type-classes. Since then, the Haskell kind-system has been enriched with functional dependencies, GADTs, type families, kind polymorphism, and type-promotion.

Together, these extensions provide the building blocks for type-level programming in Haskell.

Returning to our earlier example, let's write a type-level function that computes type-level numbers. Since we have lists with types that encode their size, we can write an append function that encodes the appended list size in the return type:

```
vappend :: VecD e n -> VecD e m -> VecD e (Add n m)
vappend NilD        l   = l
vappend (ConsD x xs) ys  = ConsD x (vappend xs ys)
```

The Add is a type level function that adds two type-level numbers. We can express this as a type family:

```
type family Add (n :: Nat) (m :: Nat) :: Nat
```

The kind signature constrains the types in an analogous way to regular type signatures constraining terms. Instances of the family use pattern matching on types instead of pattern matching on data-constructors:

```
type instance Add ZeroD m = m
type instance Add (SuccD n) m = SuccD (Add n m)
```

Let's append two `Vec` data types:

```
xs = vappend (ConsD 3 (ConsD 5 NilD))
             (ConsD 7 NilD)

-- xs :: VecD Integer ('SuccD ('SuccD ('SuccD 'ZeroD)))
```

The `Add` type-function calculated the combined vector size. The `Add` function demonstrates the humble beginnings of type-level arithmetic (the arithmetic that is executed during the type-checking phase). This is a well explored topic in the Haskell community (`https://wiki.haskell.org/Type_arithmetic`).

Promoting term-level programs to type-level

At this point in our discussion, type-level programming still lacks the expressiveness of term-level programming. In particular, we don't have these language features on type-level: case expressions, anonymous functions, partially applied functions, and let expressions.

This changed in 2014 (refer to *Promoting functions to type families* in Haskell, by *Eisenberg and Stolarek*) when these features were added to type-level using template-programming techniques. These techniques culminated in the singletons library (`https://hackage.haskell.org/package/singletons`).

By further extending the type-level in these ways, we can write (more and more) code at term-level that can automatically be promoted to type-level. Put another way, we can write code that operates at both term and type levels.

Closed type families

In 2014, the type-family picture was completed (for now) with the addition of closed type families, where all instances of the type family are declared in one place. This restriction brings the benefit of type-inference, which is lost with open type families (refer to *Promoting Functions to Type Families*, 2014, Eisenberg et al):

```
-- {-# LANGUAGE TypeFamilies #-}
-- closed-type family
type family IsZero (n::Nat) :: Bool where
  IsZero 'ZeroD = 'True
  IsZero ('SuccD n) = 'False
```

The `IsZero` is a type-level function that matches pattern against types.

The history of type-level programming in Haskell

The following table illustrates the history of type-level programming in Haskell:

Year	Language extension	Advance in Type-level programming
1997	MultiParamTypeClasses	Basis for Type functions
2000	FunctionalDependencies	Type-level functions (relational-style)
2003	GADTs	Type refinement
2005	TypeFamilies — Associated Types	Type-level functions, type-class bound
2008	TypeFamilies — top-level and associated types unified	Type-level functions, top level and type-class bound
2012	PolyKinds — Kind polymorphism	Type-level polymorphism
2012	DataKinds — Type Promotion	Type-level datatypes, type-safety
2014	TypeFamilies — Closed-type families	Type families with type-inference
2014	Singletons library	Type-level: case, let, lambdas, currying

Type-level and generic programming

Type-level programming can be considered to be another pattern of generic programming. Furthermore, it interacts with two other types of generic programming:

- As we saw in the examples of this chapter, type-level programming techniques are useful in implementing datatype generic programming. Also, there are some generic programs that can be written at type-level instead of term-level.

- We also saw that template metaprogramming is used in type-level programming and datatype generic programming.

While template metaprogramming, datatype generic programming, and type-level programming all happen at different levels of abstraction and different phases of program execution, they also weave together and push each other forward.

Dependently-typed programming

Dependently-typed programming refers to type-level programming where prior data types determine the types of subsequent values.

By doing type-checking computations, dependently-typed programming style allows for more nuanced type definitions. For instance, instead of defining a type of "list of numbers", we might go further with the dependently-typed "list of numbers of size n" or "list of distinct strings".

For example, it is easy to implement `sprintf` in an untyped way, but this function is notoriously difficult to implement in a type-safe language because the return type depends on the value of the format string:

```
sprintf "%s"  :: String -> String
sprintf "%d"  :: Int -> String
```

To do this in Haskell, we require dependently-typed programming. Let's explore a simplified example (refer to *Fun with Types*, *Kiselyov et al*, for more information). First, we create an embedded language for format strings:

```
data L       -- literals e.g. "hello"
data V val   -- values e.g. (V Int) or (V String)

-- F: format
data F t where
    Lit :: String -> F L
    Val ::  (val -> String) -> F (V val)
```

The GADT `F` unifies `L` and `V` into a type to express print formats. The `Lit` function takes a string and returns the format for literal strings `F L`. The `Val` value takes a polymorphic "to string" function and returns the format for typed values `F (V val)`.

Next, we create a type-family to generate the appropriate type signatures for our `sprintf` function:

```
type family SPrintf f
type instance SPrintf L        = String
type instance SPrintf (V val)  = val -> String
```

We can now write our generic `sprintf` function with the return types computed by the type function `SPrintf`:

```
sprintf :: F f -> SPrintf f
sprintf (Lit str)   = str
sprintf (Val show') = \x -> (show' x)
```

We can use this to print various types:

```
sprintf (Lit "hello")

-- sprintf :: SPrintf (V Float)
sprintf (Val (show::Float -> String)) 1.2

-- sprintf :: SPrintf (V String)
sprintf (Val (show::String -> String)) "hello"
```

For `Val` values, `sprintf` returns a function of the appropriate type, while `sprintf` (`Lit "hello"`) simply returns a string value. This is generic programming in the dependently-typed style.

Haskell and dependently-typed programming

Dependently-typed programming has been around even before Haskell and has evolved alongside it. Over time, both sides have started to influence each other (for example, GADTs are an import from that paradigm.)

There is still a clear line drawn between term-level and type-level in Haskell, or, as Conor McBride puts it, "the barrier represented by `::` has not been broken".

However, with each new kind-level language extension, the line is getting more blurred. Leap by lurch, Haskell is reaching towards dependently-typed programming.

As we extend the language, the inference typically suffers. We need to add more and more type signatures to annotate our code. Dependently-typed languages, on the other hand, are built on a foundation of strongly-typed and inferable type-level programming. But there are some benefits of accessing dependently-typed programming through Haskell, as described in *Giving Haskell a Promotion*, by *Yorgey et al*, in 2012:

> "*Full-spectrum dependently-typed languages like Coq or Agda are more expressive still. But these languages are in some ways too powerful: they have a high barrier to entry both for programmers and implementers. Instead, we start from the other end: we carefully extend a state-of-the-art functional programming language with features that appear in dependently-typed languages. Our primary audience is the community of Designers and Implementers of Typed Languages, to whom we offer a big increase in expressive power for a very modest cost in terms of intellectual and implementation complexity.*"

While the mainstream programming communities wrestle with marrying OOP and FP, the Haskell community is exploring the synthesis of FP and dependently-typed programming.

Summary

This chapter reviewed patterns of kind abstraction. In doing so, we encountered the key ways in which the kind-system has been extended since Haskell 98.

We started with associated type synonyms, which we then placed in the broader context of type families. Next, we explored kind polymorphism and type promotion and found that these kind-system enrichments together raise Haskell to a capable type-level programming language.

As the kind system becomes more powerful, the line between type and term-level programming becomes fainter. Yet, we saw that kinds remain second-class citizens of Haskell.

We concluded with a discussion of dependently-typed programming and how the Haskell language continues to reach more deeply into this paradigm.

Epilogue

Christopher Alexander, the father of design patterns in architecture, describes a pattern as a solution "you can use a million times over, without ever doing it the same way twice". In our case, we've looked at high level design patterns to lower level idioms and everything in between.

This book has been the story of modern Haskell through the lens of patterns. We saw that Haskell is evolving along several fronts:

- libraries that deal with specific domains for example, Iteratee Streaming IO libraries, Foldable, Traversable, the Lens library
- the type system language extensions for example, Rank-n types, existential and phantom types, GADTs, functional dependencies
- Generic programming techniques: from template meta-programming to Origami datatype generic programming
- the kind system language extensions: type functions, kind polymorphism, type promotion

As this evolution occurs, the real gems that are being forged are the underlying idioms and patterns, because they will ultimately outlast the Haskell language itself:

> *"One day, Haskell will be no more than a distant memory. But we believe that, when that day comes, the ideas and techniques that it nurtured will prove to have been of enduring value through their influence on languages of the future."*

> *–History of Haskell, Hudak et al*

Index

Symbols

A

actions
 sequencing, with applicative 51
 sequencing, with monad 51
ad hoc datatype genericity 104
Ad-hoc polymorphism
 about 9
 alternation-based 10
 alternation-based, versus class-based 11
 class-based 10
 dispatch 11-13
 parametric, unifying 13
 visitor pattern 11-13
algebraic types
 about 7
 and pattern matching 7
 recursive types 8
applicative
 actions, sequencing with 51
 monad as 50
applicative functor 45, 46
arrows
 about 59, 60
 implementing 60, 61
 Kleisli arrows 63, 64
 monad arrows 63, 64
 need for 64, 65
 operators 61-63
associated type synonyms
 about 128-130
 data families 132
 functional dependencies, using 128, 129

 type (synonym) families 130, 131
 versus functional dependencies 130

B

bind chain
 and monad 51, 52

C

currying functions
 about 3
 and composability 4
 decoupling with 5

D

datatype-generic function
 writing 109-111
data types abstraction
 about 89
 existential quantification 90, 91
 Generalized algebraic data
  types (GADTs) 94
 Phantom types 91-93
 universal quantification 89, 90
dependently-typed programming
 about 141
 and Haskell 142
domain specific language (DSL) 91
dynamic types 96, 97

E

existential quantification 90, 91
extensible-effects
 URL 56

F

foldable 71-73
folding
  over lists 68
  with monadic functions 68, 69
  with monoids 69, 70
foldr function
  URL 68
function
  about 13
  behavior, decoupling 15
  code, modularizing 15
functional dependencies
  versus associated type synonyms 130
functional programming (FP) 1
Functional Reactive Programming
  URL 65
function composition 43
function types abstraction
  RankNTypes 88, 89
functor
  about 43-45
  applicative functor 45, 46
  monad as 49
Functor-Applicative-Monad
  URL 50

G

Generalized algebraic data types (GADTs)
  dynamic types 96, 97
  heterogeneous lists 98
  Typecase pattern 95, 96
generic fold 115, 116
generic map 114
generic programming, patterns
  about 104
  datatype generic programming 106
  functions 104
  meta-programming 105
  polymorphic types and functions 104
  type laws 105, 106
generic unfold 116
GHC.Generics 112

Glasgow Haskell Compiler (GHC)
  about 105
  URL 130

H

handle-based I/O
  about 27
  URL 27
Haskell
  about 87, 98
  community, URL 139
  modernizing 78, 79
heterogeneous lists
  existentials used 98
  GADTs used 99
higher-kinded polymorphism
  about 127
higher-order functions (HOF)
  about 2, 14
  as first-class citizens 2
  composing 3
  currying 3
higher-order kinds 126
Hinze
  URL 97

I

imperative I/O 25-27
IO
  as applicative 25
  as functor 25
IO monad 24, 25
Iteratee I/O
  about 35-37
  enumeratees, generated 40, 41
  enumerator 39
  enumerators, generated 40, 41
  I/O styles 42
  Iteratee 37-39
  iteratees, generated 40, 41
  libraries 41
  libraries, URL 41
  URL 36

Thank you for buying
Haskell Design Patterns

About Packt Publishing

Packt, pronounced 'packed', published its first book, *Mastering phpMyAdmin for Effective MySQL Management*, in April 2004, and subsequently continued to specialize in publishing highly focused books on specific technologies and solutions.

Our books and publications share the experiences of your fellow IT professionals in adapting and customizing today's systems, applications, and frameworks. Our solution-based books give you the knowledge and power to customize the software and technologies you're using to get the job done. Packt books are more specific and less general than the IT books you have seen in the past. Our unique business model allows us to bring you more focused information, giving you more of what you need to know, and less of what you don't.

Packt is a modern yet unique publishing company that focuses on producing quality, cutting-edge books for communities of developers, administrators, and newbies alike. For more information, please visit our website at www.packtpub.com.

About Packt Open Source

In 2010, Packt launched two new brands, Packt Open Source and Packt Enterprise, in order to continue its focus on specialization. This book is part of the Packt Open Source brand, home to books published on software built around open source licenses, and offering information to anybody from advanced developers to budding web designers. The Open Source brand also runs Packt's Open Source Royalty Scheme, by which Packt gives a royalty to each open source project about whose software a book is sold.

Writing for Packt

We welcome all inquiries from people who are interested in authoring. Book proposals should be sent to author@packtpub.com. If your book idea is still at an early stage and you would like to discuss it first before writing a formal book proposal, then please contact us; one of our commissioning editors will get in touch with you.

We're not just looking for published authors; if you have strong technical skills but no writing experience, our experienced editors can help you develop a writing career, or simply get some additional reward for your expertise.

Learning Haskell Data Analysis

ISBN: 978-1-78439-470-7   Paperback: 198 pages

Analyze, manipulate, and process datasets of varying sizes efficiently using Haskell

1. Create portable databases using SQLite3 and use these databases to quickly pull large amounts of data into your Haskell programs.

2. Visualize data using EasyPlot and create publication-ready charts.

3. An easy-to-follow guide to analyze real-world data using the most commonly used statistical techniques.

Mastering Machine Learning with scikit-learn

ISBN: 978-1-78398-836-5   Paperback: 238 pages

Apply effective learning algorithms to real-world problems using scikit-learn

1. Design and troubleshoot machine learning systems for common tasks including regression, classification, and clustering.

2. Acquaint yourself with popular machine learning algorithms, including decision trees, logistic regression, and support vector machines.

3. A practical example-based guide to help you gain expertise in implementing and evaluating machine learning systems using scikit-learn.

Please check **www.PacktPub.com** for information on our titles

Haskell Data Analysis Cookbook

ISBN: 978-1-78328-633-1   Paperback: 334 pages

Explore intuitive data analysis techniques and powerful machine learning methods using over 130 practical recipes

1. A practical and concise guide to using Haskell when getting to grips with data analysis.

2. Recipes for every stage of data analysis, from collection to visualization.

3. In-depth examples demonstrating various tools, solutions and techniques.

Haskell Financial Data Modeling and Predictive Analytics

ISBN: 978-1-78216-943-7   Paperback: 112 pages

Get an in-depth analysis of financial time series from the perspective of a functional programmer

1. Understand the foundations of financial stochastic processes.

2. Build robust models quickly and efficiently.

3. Tackle the complexity of parallel programming.

Please check **www.PacktPub.com** for information on our titles

93174029R00093

Made in the USA
Middletown, DE
12 October 2018